AGRONOMY FOR DEVELOPMENT

Over the last decade there has been renewed interest in food security and the state of the global food system. Population growth, climate change and food price spikes have combined to focus new attention on the technologies and institutions that underpin the production and consumption of food that is varied, nutritious and safe.

Knowledge politics within development-oriented agronomy set the stage for some models of agricultural development to be favoured over others, with very real implications for the food security and wellbeing of many millions of people. *Agronomy for Development* demonstrates how the analysis of knowledge politics can shed valuable new light on current debates about agricultural development and food security. Using bio-physical and social sciences perspectives to address the political economy of the production and use of knowledge in development, this edited collection reflects on the changing politics of knowledge within the field of agronomy and the ways in which these politics feed and reflect the interests of a broad set of actors.

This book is aimed at professionals working in agricultural research as well as students and practitioners of agricultural, rural and international development.

James Sumberg is Research Fellow in the Rural Futures research cluster at the Institute of Development Studies (IDS) and a member of the STEPS Centre, University of Sussex, UK.

Pathways to Sustainability Series
Series Editors:
Ian Scoones and Andy Stirling
STEPS Centre at the University of Sussex

Editorial Advisory Board:
Steve Bass, Wiebe E. Bijker, Victor Galaz, Wenzel Geissler, Katherine
Homewood, Sheila Jasanoff, Melissa Leach, Colin McInnes, Suman Sahai,
Andrew Scott

This book series addresses core challenges around linking science and technology
and environmental sustainability with poverty reduction and social justice. It is
based on the work of the Social, Technological and Environmental Pathways to
Sustainability (STEPS) Centre, a major investment of the UK Economic and Social
Research Council (ESRC). The STEPS Centre brings together researchers at the
Institute of Development Studies (IDS) and SPRU (Science Policy Research Unit)
at the University of Sussex with a set of partner institutions in Africa, Asia and
Latin America.

Titles in this series include:

Gender Equality and Sustainable Development
Edited by Melissa Leach

Adapting to Climate Uncertainty in African Agriculture
Narratives and knowledge politics
Stephen Whitfield

One Health
Science, politics and zoonotic disease in Africa
Edited by Kevin Bardosh

Grassroots Innovation Movements
Adrian Smith, Mariano Fressoli, Dinesh Abrol, Elisa Around and Adrian Ely

Agronomy for Development
The politics of knowledge in agricultural research
James Sumberg

AGRONOMY FOR DEVELOPMENT

The Politics of Knowledge in Agricultural Research

Edited by James Sumberg

Routledge
Taylor & Francis Group

LONDON AND NEW YORK

earthscan
from Routledge

First published 2017
by Routledge
2 Park Square, Milton Park, Abingdon, Oxon OX14 4RN

and by Routledge
711 Third Avenue, New York, NY 10017

Routledge is an imprint of the Taylor & Francis Group, an informa business

British Library Cataloguing-in-Publication Data
A catalogue record for this book is available from the British Library

Library of Congress Cataloging-in-Publication Data
A catalog record for this book has been requested

ISBN: 978-1-138-24027-8 (hbk)
ISBN: 978-1-138-24031-5 (pbk)
ISBN: 978-1-315-28405-7 (ebk)

Typeset in Bembo
by Saxon Graphics Ltd, Derby

CONTENTS

ABBREVIATIONS

AATF	African Agricultural Technology Foundation
ACB	African Center for Biodiversity
ACF	Advocacy Coalition Framework
ACTS	African Centre for Technology Studies
ADB	Agricultural Development Bank (Ghana)
AESD	Agricultural Engineering Services Directorate
AFSA	African Forum for Food Sovereignty
AGRA	Alliance for a Green Revolution for Africa
AKIS	Agricultural Knowledge and Information System
AML	African Model Law
AMSEC	Agricultural Mechanisation Service Centre
AR4D	Agricultural Research for Development
ARIPO	African Regional Intellectual Property Organization
ASA	Agricultural Seed Agency
ASARECA	Association for Strengthening Agricultural Research in Eastern and Central Africa
AWD	Alternate Wetting and Drying
BMGF	Bill & Melinda Gates Foundation
CA	Conservation Agriculture
CAADP	Comprehensive *African Agriculture Development Programme*
CEBIB	Centre for Biotechnology and Bioinformatics
CEO	Chief Executive Officer
CFDT	Compagnie Française pour le Développement des Fibres Textiles
CGIAR	Consultative Group on International Agricultural Research
CIMMYT	International Maize and Wheat Improvement Center
CIP	International Potato Centre
COMESA	Common Market for Eastern and Southern Africa
CRP	CGIAR Research Programme

CRS	Catholic Relief Services
CSA	Climate Smart Agriculture
CSO	Civil Society Organisation
DFID	UK Department for International Development
DSIP	Development Strategy and Investment Plan
DVM	Decentralized Vine Multiplication
EFSA	European Food Safety Authority
EU	European Union
FAO	Food and Agriculture Organisation of the United Nations
GADCO	Global Agri-Development Cooperation
GAFSP	Global Agriculture and Food Security Program
GCARD	Global Conference on Agricultural Research for Development
GEF	Global Environment Facility
GIDA	Ghana Irrigation Development Authority
GM	Genetic Modification
GMO	Genetically Modified Organism
HKI	Helen Keller International
IAASTD	International Assessment of Agricultural Knowledge, Science and Technology for Development
IAD	Institutional Analysis and Development
IDO	Intermediate Development Outcome
IFAD	International Fund for Agricultural Development
IFPRI	International Food Policy Research Institute
IITA	International Institute of Tropical Agriculture
INGO	International Non-Governmental Organisation
IPC	International Planning Committee for Food Sovereignty
IPG	International Public Good
ISAAA	International Service for the Acquisition of Agri-biotech Applications
ISPC	Independent Science and Partnership Council
KARI	Kenya Agricultural Research Institute
KEGCO	Kenya GMO Concern Group
KfW	German Development Bank
LDC	Least Developed Country
MAAIF	Ministry of Agriculture, Animal Industry and Fisheries (Uganda)
MOFA	Ministry of Food and Agriculture (Ghana)
MP	Member of Parliament
MST	Movimiento de Trabajadores Rurales Sin Tierra
NAADS	National Agricultural Advisory Services (Uganda)
NACOSTI	National Commission for Science, Technology and Innovation (Kenya)
NARO	National Agricultural Research Organization (Uganda)
NBA	National Biosafety Authority
NCST	National Council for Science and Technology
NEPAD	New Partnership for Africa's Development

NERICA	New Rice for Africa
NGO	Non-Governmental Organisation
NMRP	National Maize Research Program
NPM	New Public Management
NWO	Dutch Organisation for Scientific Research
OECD	Organisation for Economic Co-operation and Development
OFAB	Open Forum on Agricultural Biotechnology
OFSP	Orange-fleshed Sweet Potato
OHVN	Office de la Haute Vallée du Niger
OPV	Open Pollinated Variety
PBRs	Plant Breeders' Rights
PBS	Programme on Biosafety
PEAP	Poverty Eradication Action Plan
PMA	Plan for Modernisation Agriculture
PPP	Public Private Partnership
PVP	Plant Variety Protection
QDS	Quality Declared Seed
QPM	Quality Protein Maize
R&D	Research and Development
SES	Social-ecological System
SLO	System Level Outcome
SPHI	Sweet Potato Profit and Health Partnership
SRF	Strategic Results Framework
SRI	System of Rice Intensification
SRT	Strategic Research Theme
SSA	Sub-Saharan Africa
SSC	South-South Cooperation
STEPS	Social, Technological and Environmental Pathways to Sustainability (STEPS Centre)
STS	Science and Technology Studies
T&V	Training and Visit
TAAP	Tanzanian Agricultural Agribusiness Program
TABIO	Tanzania Alliance for Biodiversity
TAHEA	Tanzanian Home Economics Association
TOSCI	Tanzania Official Seed Certification Institute
UK	United Kingdom
UN	United Nations
UNEP	United Nations Environment Programme
US	United States of America
US$	United States dollar
USAID	US Agency for International Development
VISTA	Viable Technologies for Sweet Potato in Africa
WEMA	Water Efficient Maize for Africa
WHO	World Health Organization
WTO	World Trade Organization

CONTRIBUTORS

Kojo Amanor is Professor at the Institute of African Studies, University of Ghana, Legon, Ghana.

Jens A. Andersson is a Rural Development Sociologist in the Sustainable Intensification Programme of the International Maize and Wheat Improvement Center, Mexico.

Regina Birner is Professor of Social and Institutional Change in Agricultural Development at the Hans-Ruthenberg-Institute, University of Hohenheim, Stuttgart, Germany.

James A. Fraser is a Lecturer at Lancaster Environment Centre, Lancaster University, UK.

Ken E. Giller is Professor in the Plant Production Systems Group, Wageningen University, The Netherlands.

Dominic Glover is a Research Fellow at the Institute of Development Studies, UK.

Chris Huggins is Assistant Professor at the School of International Development and Global Studies, University of Ottawa.

Paul Kibwika is Associate Professor at the Department of Agricultural Extension and Innovation Studies, Makerere University, Kampala.

Laurens Klerkx is Associate Professor in the Knowledge, Technology and Innovation Group, Wageningen University, The Netherlands.

Cees Leeuwis is Professor of Knowledge, Technology and Innovation, Wageningen University, The Netherlands.

Harro Maat is a Lecturer in the Knowledge, Technology & Innovation Group, Wageningen University, The Netherlands.

Margaret N. Mangheni is Associate Professor at the Department of Agricultural Extension and Innovation Studies, Makerere University, Kampala.

Frank B. Matsiko is Senior Lecturer in the Department of Agricultural Extension and Innovation Studies, Makerere University, Kampala.

William G. Moseley is Professor of Geography, and Director of African Studies, Macalester College, USA.

Sheila Rao is a Research Consultant and PhD Candidate at the Department of Anthropology and Sociology, Carleton University.

Patience B. Rwamigisa is Head of the Department of Agricultural Extension and Skills Management at the Ministry of Agriculture, Animal Industry and Fisheries of the Republic of Uganda, Uganda.

Marc Schut is a Social Scientist working with the International Institute of Tropical Agriculture (IITA) and Wageningen University, based in Rwanda.

James Sumberg is a Research Fellow at the Institute of Development Studies, UK.

John Thompson is a Research Fellow at the Institute of Development Studies, UK.

Jean-Philippe Venot is a Senior Researcher at the French National Research Institute for Sustainable Development, based in Phnom Penh, Cambodia. He is also affiliated to the Water Resources Management Group at Wageningen University.

Ola T. Westengen is Associate Professor at the Department of International Environment and Development Studies (Noragric), Norwegian University of Life Sciences, Norway.

Stephen Whitfield is Lecturer in Climate Change and Food Security at the University of Leeds, UK.

ACKNOWLEDGEMENTS

This book is the product of an international conference held in February 2016 at the Institute of Development Studies (IDS) and sponsored by the STEPS Centre at the University of Sussex, the International Maize and Wheat Improvement Center (CIMMYT) through the CGIAR research programmes on MAIZE and WHEAT and Wageningen University. Their support is gratefully acknowledged, as are the contributions of all those who participated in the conference.

1

KNOWLEDGE POLITICS IN DEVELOPMENT-ORIENTED AGRONOMY

Jens A. Andersson and James Sumberg

Introduction

Over the last two decades there has been renewed concern about food security and the state of the global food system. After a long period in which agricultural research and development issues were largely neglected by national governments and the international community (Adenle *et al.* 2013), population growth, climate change, food price spikes, food safety scares and food-related disease have brought debates about food production and consumption back into the public arena (Ingram *et al.* 2010; HLPE 2011; UK Foresight 2011). For example, the World Bank largely ignored the central role of agriculture in rural development throughout much of the late 1980s and 1990s: its *World Development Report 2008: Agriculture for Development* (World Bank 2007) was the first time in 25 years the Bank's flagship publication was devoted to agriculture. With this renewed concern has come a rekindled appreciation of the agricultural sciences, which are increasingly understood to have a major role to play in addressing these challenges. Bringing together the biological, technical and human factors in agricultural production, agronomy is at the very centre of efforts to sustainably enhance agricultural productivity.

Yet we suggest that the discipline of agronomy is also in a state of transition, and at the heart of this transition are questions such as 'What agronomy?', 'Whose agronomy?' and 'Whose agronomy counts?'. While such questions are being asked about agronomy in general, our focus is specifically on agronomy as a component of rural development in the global South. We call this 'development-oriented agronomy', which is itself a broad church. It includes a myriad of sub-specialties relating to crop and pasture production, including but not limited to crop and soil management, crop improvement, irrigation, crop-livestock integration, farm and farming systems analysis. It is undertaken in and funded by a range of public and private sector organisations, in the South and in the North. A defining characteristic of development-oriented agronomy is that it is motivated by a desire to contribute to one or more development challenges, like food and nutrition security, human welfare and wellbeing, or environmental sustainability.

Development-oriented agronomy is no stranger to debate and contestation. For example, there has been a protracted debate about the role of agronomic technology and management practices in the Green Revolution in Asia, and their poverty, equity and environmental effects – although not much of this debate has appeared in mainstream agronomy journals (Griffin 1974; Conway 1999; Evenson and Gollin 2003; Orr 2012; Patel 2013). Recent years have also seen claims and counter claims, as well as contradictory conclusions, endorsements, declarations and recommendations about, for example, the performance and potential value of genetically modified (GM) crops, the System of Rice Intensification (SRI), Conservation Agriculture (CA), organic farming's ability to feed the world, the Savory holistic grazing method, New Rice for Africa (NERICA), Climate Smart Agriculture (CSA), Sustainable Intensification and so on. Such contestations appear to be an increasingly common feature of agronomy as a scientific discipline, and for this reason they should not be dismissed as aberrations, sideshows or signs of a failing discipline. On the contrary, as argued in *Contested Agronomy: Agricultural Research in a Changing World* (Sumberg and Thompson 2012), they are worthy of serious analysis for what they can tell us about the politics that underpin and orient the production and promotion of development-oriented agronomic knowledge.

In the next sections of this chapter we explore the notion of development-oriented agronomy in more depth. We then briefly summarise the original contested agronomy argument. Following this, the chapter focuses on the notion of knowledge politics, which sits at the core of contested agronomy. The last section links these arguments to subsequent chapters in this volume.

Development-oriented agronomy

Development-oriented agronomy as we know it today is firmly rooted in the European colonial era (Ross 2014). The ecosystems, crops, soils and farming practices that colonial officers encountered in Africa, Asia and Latin America – and the drive to use or to modify these to fuel European economic expansion – framed the development of the new sub-speciality of 'tropical agronomy'. But aside from a focus on tropical areas and the close association with the authority of the colonial project, as a field of knowledge creation, there was, initially, little inherent coherence to tropical agronomy.

The shift from tropical agronomy to development-oriented agronomy took place over a number of years, having started already in the colonial era, for example in the growing pre-occupation with the modernisation of indigenous agricultural practices. It reflected the change from colonial to independent administrations; the development of national agricultural research organisations and capabilities; a greater emphasis on food crops and rural poverty alleviation; and more generally, the emergence of the 'developmental state' (Leftwich 1995). In parallel, as colonial configurations transitioned to the new institutions and international relations of foreign aid, technical and development assistance, development-oriented agronomy took on a stronger international dimension with, for example, the establishment of the UN Food and Agriculture Organisation (FAO), the Consultative Group on International Agricultural Research (CGIAR) and regional research organisations, as well as networks of students, relations and partnerships linking agricultural universities, research institutions and projects in the South and the North.

If tropical agronomy lacked coherence as a field of knowledge creation, a similar charge can also be levelled at development-oriented agronomy. A variety of understandings of and approaches to development-oriented agronomy are evident, as exemplified by different definitions and research traditions. For example, the Oxford English Dictionary (2012) defines agronomy as 'the practice or (now chiefly) the science of crop production and soil management'. The American Society of Agronomy's *Agronomy Journal* frames agronomy essentially as technical science that has little overt concern with the social, economic or political relations within which agriculture and agronomic research are undertaken. Specifically, *Agronomy Journal* publishes:

> articles relating [...] soil-plant relationships; crop science; soil science; biometry; crop, soil, pasture, and range management; crop, forage, and pasture production and utilization; turfgrass; agroclimatology; agronomic modeling; statistics; production agriculture; and computer software.[1]

A somewhat broader perspective is apparent in the definition of *agronomie* given in *Larousse agricole*:

> *Ensemble des sciences nécessaires à la compréhension de l'agriculture et des techniques utiles à sa pratique. Au sens strict, l'agronomie est l'étude scientifique des relations entre les plantes cultivées, le milieu (sol, climat) et les techniques agricoles. Dans un sens plus large, elle comprend aussi l'ensemble des sciences et des techniques relatives à l'élevage, à la sylviculture, au génie rural. Enfin, l'économie, la sociologie, la comptabilité et la gestion de l'exploitation agricole sont aujourd'hui considérées comme des sciences nécessaires à la compréhension des techniques.*

> [All the sciences necessary to understand agriculture and the techniques used to practice it. Strictly speaking, agronomy is the scientific study of the relationships between cultivated plants, the environment (soil, climate) and agricultural techniques. In a much broader sense, it also includes all sciences and technologies related to livestock (breeding), forestry and rural engineering. Finally, economics, sociology, accounting and management of the farm are nowadays also considered necessary for understanding technology.]

> (*Mazoyer* et al. 2002)

One might conclude that the Francophone tradition of agronomy and agronomic training is inherently more holistic, situated and systems-oriented. However, it would probably be a mistake to draw too sharp a distinction between Anglophone and Francophone traditions in development-oriented agronomy. For instance, both made important contributions to the farming systems research movement of the 1980s and 1990s (Fresco 1984; Jouve 1986; Brossier 1987; Collinson 2000), which can be seen as an attempt to broaden and situate development-oriented agronomy (although at least on the Anglophone side, largely led by economists). A desire to broaden and situate agronomy can also be discerned in the sustainable agriculture and agroecology movements (Wezel *et al.* 2009), and in the interest within the CGIAR and elsewhere in so-called 'agricultural research for development' (AR4D) (Virchow and von Braun 2001; von Kaufmann 2007; cf. Coe *et al.* 2014).

There is a line of argument, which suggests that every farmer could be considered to be an agronomist (Allan 1965; Richards 1985), and to some degree an experimental agronomist (Richards 1986; Sumberg and Okali 1997). However, as important as they may be, our central interest is not 'people's agronomy', 'folk agronomy' or farmers' experiments. Rather, our focus is on formal research within the discipline of agronomy, including those activities that generate new agronomic knowledge, as well as the use of that knowledge to promote new practices.

The contested agronomy argument

As set out by Sumberg, Thompson and Woodhouse (2012, 2013) the contested agronomy argument has four main elements:

1. Over the last four decades the context within which agronomic research takes place has been transformed fundamentally, with the most important changes being the rise of (a) the neoliberal project; (b) the environmental agenda; and (c) the participation agenda.
2. As a result, the long-standing unity of purpose between the state on the one hand, and the agronomic research establishment on the other, has been undermined. Agronomy has ceased to be the handmaiden of the state, with important implications for how development-oriented agronomy in particular is conceived, funded, managed, implemented, evaluated and portrayed.
3. With less unity of purpose, and in the more crowded, competitive, short-term and impact-oriented funding context, development-oriented agronomy has become an altogether more contested and contentious space: the politics around agronomic knowledge is now less controlled and much more public.
4. This new knowledge politics around development-oriented agronomy is having important impacts on the discipline itself, and on its ability to address the challenge of sustainably enhancing agricultural productivity.

The environmental and participation agendas certainly helped to restructure the language and re-orient the gaze of development-oriented agronomy. For example, today when words like 'agriculture', 'farming systems' and 'intensification' appear without being preceded by 'sustainable', it is as if a sacred norm has been publicly and dangerously flouted. Also, the propositions that farmers are knowledgeable, that agronomic research should be client- or demand-driven, and that potential users of technology should have a role in technology development, are now commonly accepted among agronomists. The extent to which the environmental and participation agenda have really changed the framings, priorities, methods or outputs of development-oriented agronomy is open to question, particularly in the face of the panic around 'feeding 9 billion people by 2050' and the resulting narrow focus on the use of technology to increase production and close yield gaps. Nevertheless, there can be little doubt that new spaces have been opened up for the public contestation of agronomy's goals, priorities, methods, findings and recommendations.

At the same time, the effects of the neoliberal project on development-oriented agronomy have been and continue to be very real. The goal of shrinking the state and strengthening accountability led to the embedding of New Public Management (NPM) principles in international and national agricultural research institutions (Dunleavy and Hood 1994; Manning 2001). These principles include performance auditing and measurement, privatisation, competition, strategic planning and management, and public–private partnership (Gruening 2001). For publicly funded agricultural research this has meant a fundamental change in the basis on which research programmes are conceived, funds allocated and investments evaluated.

The influence of NPM can be seen in the emphasis that funders of agricultural research now place on developing results frameworks, elaborating targets, identifying quick wins, meeting delivery schedules, theorising impact pathways, and demonstrating 'value for money', and ultimately, impact at scale. The CGIAR's *Strategy and Results Framework 2016–2025* with its vision, mission, three strategic goals or system-level outcomes (SLOs), ten Intermediate Development Outcomes (IDOs) and 32 sub-IDOs is a case in point,[2] but it is certainly not unique. While these instruments have become an accepted part of the agricultural research landscape, we know of no research that investigates their effects on research processes or outputs. Sumberg *et al.* (2013) hypothesised that their imposition would likely favour downstream over upstream and short-term over long-term research, which would be at odds with the rhetorical focus on sustainability.

The need to demonstrate impact at scale creates strong incentives for individuals and organisations at all levels – from field-based research agronomists, through programmes and partnerships, to bi- and multi-lateral research funders – to draw attention to, and defend their particular contributions, their successes. It is ironic that in an era of 'evidence-based policy making' and 'rigorous impact evaluation', the success story has become a format of choice in the struggle to convince funders, politicians and the public of the benefits of development-oriented agricultural research (e.g. Wiggins 2009; Spielman and Pandya-Lorch 2009; cf. Sumberg, Irving, *et al.* 2012).

It is in this context that a focus on knowledge politics within development-oriented agronomy becomes particularly important.

Knowledge politics in development-oriented agronomy

In the social sciences, the idea that knowledge and evidence are inseparable from power and politics is now widely accepted. Robert Chambers' simple question 'Whose knowledge?' (Chambers 1983) reminds us of the existence of different knowledges, and the inherent limitations of thinking in terms of a single objective truth or reality. The political angle comes into play as individuals and groups selectively generate and/or use knowledge to establish, maintain or enhance their vested interests. The common and logical presumption in much of the literature on knowledge politics is that powerful actors are best placed to do this successfully.

Contested Agronomy: Agricultural Research in a Changing World (Sumberg and Thompson 2012), primarily sought to draw attention to the rise of more explicit knowledge politics within and around the field of agronomy. It sought to open up the analysis of agronomic knowledge production, appropriation and application in terms of different interests, networks, epistemic communities and asymmetric power relations (Andersson and Giller 2012; Brooks and Johnson-Beebout 2012; Sumberg, Thompson and Woodhouse 2012). In so doing it brought to the fore the role of the agronomist, but perhaps more importantly, the role of agronomic research institutes, as political actors. Focusing on specific contestations and alliances among agronomists, and between agronomists, policy makers, practitioners and other development actors, provides an obvious entry-point for analysis of the politics of development-oriented agronomic research. This focus highlights important questions such as: 'Who are the powerful actors?', 'What are their vested interests?' and 'How is knowledge created and mobilised within epistemic communities to construct and promote particular framings and narratives in support of these interests?'

If taken seriously, a knowledge politics perspective means that agronomy can no longer be regarded as the preserve of a particular category of scientists focusing on, for example, soil management or crop production. After all, the politics of agronomic knowledge production supports and reflects the interests of a much wider set of actors, from national and international public-sector research organisations to multinational agri-food corporations, food sovereignty campaigners, farmers and consumers. The analysis of knowledge politics should help explain why particular agricultural technologies and development pathways are favoured over others (e.g. Vanloqueren and Baret 2009), with, in some cases, very real implications for peoples' food security and wellbeing, resource use and conservation. Putting a spotlight on the politics of development-oriented agronomy and its pathways from the past to the future can contribute to the opening up and enrichment of debate and deliberation about desirable futures for farming, rural economies, food systems and the environment (Leach *et al.* 2010; Thompson and Sumberg 2012). Unfortunately, what the spotlight reveals is not always pretty, nor does it necessarily lead to simple recommendations for action.

Focusing on knowledge politics, as the case studies in this volume do, is not merely an exercise in highlighting the political nature of agricultural research, or enhancing our understanding of the contestations that pervade contemporary development-oriented agronomy. A focus on knowledge politics also provides a perspective on the current and future directions of agricultural research, which ultimately influences future agricultural landscapes and food systems.

It is indisputable that farms, farming systems, agricultural systems and food systems have a past and a partially observable present. It is also true that there can be heated debate about the nature of these pasts and presents. Critically, they also have multiple possible futures. Some of these possible futures might be foreseen (e.g. by projecting well-established trends or modelling extant or emerging drivers of change); others might be imagined (either literally or by drawing on experience within an existing niche); and still others, dependent perhaps on developments yet to be imagined (much less manifest) remain masked. One or more of these futures will eventually emerge as a new present through a complex, messy, contested, stop-and-go, co-evolutionary dynamic. The combination of partially knowable past and present, and multiple possible futures, is what the STEPS Centre refers to as 'pathways' (Leach *et al.* 2010; Thompson and Sumberg 2012).

Underpinning this conception of pathways is the proposition that the world is becoming ever more complex, interconnected and unpredictable; the acknowledgement of the importance of power relations in policy and politics; and the realisation that there are multiple legitimate perspectives on sustainability. It follows that movement toward a more socially just and sustainable world must necessarily involve open and informed deliberation on possible pathways, across all aspects of the economy including farming and agriculture. An important element of this agenda is to understand why and how some pathways come to dominate over others, and how and under what conditions alternative pathways emerge. Although development-oriented agronomy may seem focused on current problems – how to control a disease, close a yield gap or use irrigation water more efficiently – its framing of research problems, methods and findings, contributes to the evolution of existing or the emergence of new pathways. Each step along the way, a decision to fund research on hybrid rice, a decision to transfer a technology from one site to another, a decision to privilege market-based approaches to extension, or a policy process that results in a relatively restrictive GMO regulation regime, is both potentially important, and deeply political. The chapters of this book highlight the contested politics around the framings, decisions, technologies and programmes that can broadly be considered as development-oriented agronomy, and which more or less successfully contribute to the emergence of the pathways of tomorrow.

Knowledge politics in agronomy: a heuristic framework

The notion of contestation is central to the analysis of knowledge politics within agronomy. Contested agronomic knowledge is disputed knowledge; it is the subject of disagreement, and it is debated and argued about. Contestation takes place around alternative, rival or contending priorities, methods, observations, interpretations and explanations.

From a knowledge politics perspective the central question is why different forms and dynamics of contestation arise and how they evolve. Here we propose a heuristic framework to analyse these questions systematically, distinguishing drivers, actors, values and interests, objects and domains of contestation.

Drivers

The politics of knowledge within development-oriented agronomy must be seen against the backdrop of long-running arguments and struggles around agrarian relations, dominant and alternative models of agriculture, and visions for rural areas and food systems. These arguments and struggles themselves reflect developments within late-modern capitalism, including changing relations of global capital, globalisation and the rise of the neoliberal agenda (e.g. McMichael 1997; Bernstein 2016).

We argue that within this historical context the dominance of the approaches and disciplines associated with New Public Management (NPM) have played a particularly important part in driving contemporary knowledge politics in publicly funded agronomic research. NPM's concerns with outcomes, accountability and value for (public) money, and the effects of placing these at the centre of agronomic research, is an unavoidable focus of attention. Is there evidence that the insistence on results frameworks, theories of change, scaling and impact pathways improves the quality of agronomic research or increases its impact? What effects do particular ways of conceiving of, monitoring and claiming success, and particularly success 'at scale', have on processes of knowledge creation and contestation?

Actors, values and interests

The key proposition is that individuals and organisations get involved in contestation in order to influence the direction of a debate, or to promote (or question) a particular approach, method, interpretation or technology. What they choose to contest and how they go about it not only reflects the incentives they face, but also their values, world views, interests and ability to get their voices heard.

For analytical purposes, actors may be grouped under seven stylised affiliations: public-sector research; policy support (e.g. the FAO); civil society; interest group (e.g. the SRI International Network and Resources Center); extension; corporation (including private sector research); and farmer. This scheme allows particular debates or contestations to be analysed in terms of 'contesting couplets' or contesting coalitions (Biggs and Smith 1998): for example, a particular debate might primarily pit private sector research against public-sector research, or public-sector research against civil society. Yet, the real world of contested agronomy is not nearly so neat, and contestation often cuts across such affiliations.

In addition, there is a degree of cross-dressing that undermines the value of these stylised affiliations – for example, when public-sector research or policy support take on the role of an interest group; when policy support organisations or an interest group presents themselves as research; or when academic journals allow themselves to be used for promotional purposes (e.g. Sumberg *et al.* 2013; Andersson *et al.* 2014). Indeed, this phenomenon of cross-dressing would appear to be a fundamental part of much of the current contestation within development-oriented agronomy, not least because it can be a powerful way to further one's interests.

Questions of interest include: When and how do the interests of different actors diverge or converge around particular issues? How are epistemic communities or discourse coalitions formed, and what values and world views underpin these? Whose success counts? Who counts success, and why?

Objects

A critically important dimension of knowledge politics in agronomy relates to the objects of contestation. Is the argument about a simple or a complex technology, a technological 'package', a set of principles, a set of options, an approach, a policy, a vision, or a particular set of institutional arrangements? Or is it about a problem-framing, the validity or appropriateness of a particular method, the interpretation of particular evidence, or some particular conclusions or recommendations? The answer to these questions sets the terms of the debate and legitimises some lines of argument and particular actors' involvement and positions, as well as some kinds of evidence over others.

Domains

We hypothesise that as the actors contesting agronomic knowledge become more diverse – that is, there is greater epistemological and social distance between them – and as the stakes associated with making and defending success claims increase, the front lines of contestation shift. Specifically, they evolve along a continuum that begins with technique and moves to framing, epistemology and values (Table 1.1). This is not a simple linear shift, but rather a progressive enlargement of the domains of contestation. Many examples of contestation in development-oriented agronomy are pursued on several fronts simultaneously. We suggest that contestation in relation to technique and framing is long-standing in agronomy. It is the expansion into the domains of epistemology and values that distinguish developments within the field over the last decade, and which are responsible for much of the 'heat' in some current debates.

We suggest that analyses of contestation and knowledge politics within development-oriented agronomy that focuses on drivers, actors and interests, objects and domains can provide new insight, and lay the basis for a more nuanced, systematic account of the changing politics of agronomic research.

TABLE 1.1 Domains of contestation

Technique, where contestation revolves around the appropriateness of experimental or research design; how a statistical model was specified; how a statistical analysis was interpreted; and the conclusions/implications drawn from an analysis

Framing, where contestation revolves around the choice of research questions; framing concepts; supporting narratives; and literature

Epistemology, where contestation revolves around the nature of evidence and knowledge; and the relative values of different kinds or sources of evidence

Values, where contestation revolves around the values and motives ascribed to particular individuals and institutions

Introduction to *Agronomy for Development*

The chapters in this volume were originally presented at the conference *Contested Agronomy: Whose Agronomy Counts?*, held at the Institute of Development Studies in February 2016.[3] The conference sought to broaden and deepen the analysis of knowledge politics in development-oriented agronomy. *Agronomy for Development* is not meant to be a comprehensive survey of contestation within the field, nor did the conference touch on all possible controversies. Rather this collection is meant to reflect the state of research on knowledge politics within agronomy and how these play out as agronomic research moves further into the realm of development.

Technology development – new crop varieties, crop and soil management practices, pest control methods etc. – is perhaps the fundamental pre-occupation of development-oriented agronomy, and a number of the chapters focus on it. If in the past the notion of technology transfer, and the functional divide between research and extension, provided agronomists with a sense of detachment and reassurance that one way or another their technologies would find their way into farmers' fields, this is clearly no longer the case. Starting with the idea that technology only becomes meaningful when it is enacted through practice, and using the concepts of inscription and affordance, Glover *et al.* (Ch. 2) take a scalpel to the black box of technology transfer. Their analysis throws important new light on the ways that agronomists as technology developers, and those who package and promote agricultural technology, try to impose their vision of how the technology should be used. It also highlights the local processes of unpacking, adaptation and hybridisation that are pervasive, and necessary to make a technology useful once it arrives in a given context. Their argument is that such analysis helps explain the recent contestation around technologies like SRI and drip irrigation. Further work along these lines will help to re-frame and reenergise debates around technological change and adoption.

The idea of technology as situated socio-technical practice also underpins the analysis by Rao and Huggins (Ch. 8) of efforts to promote orange-fleshed sweet potato (OFSP) in Tanzania. Here, technology is both process (genetic biofortification) and product (new OFSP varieties). What becomes clear is that despite the hype and high profile success stories (including the award of the 2016 World Food Prize[4]), and the fact that OFSP has been shown to deliver higher levels of vitamin A, the market-based models that are supposed to enable the scaling of both nutritional benefits and economic opportunity for women have not worked. Once the OFSPs are on the ground, these models break down in the face of well-established, non-market channels for the distribution of planting material, gender relations that result in unequal access to irrigation water, and producers' and consumers' unwillingness to displace traditional white and yellow-fleshed varieties with OFSP. Here the technology cannot be separated from the ideological commitment to a particular class of delivery mechanisms: the commitment to market-based delivery models on the part of key funders sets up a particularly power-laden, but perhaps not uncommon, form of knowledge politics.

Genetic engineering, and the GM crop varieties that it produces, have been at the centre of sustained contestation. In Kenya, as documented by Whitfield (Ch. 4), the contestation and knowledge politics around GM crops pre-dates any farm-level use of the technology. This analysis of the politics around the development of a national regulatory regime that would allow GM crop production amply illustrates the new and complex world of development-oriented agronomy. The interplay between expert knowledge ('the science') on the one hand, and uncertainty on the other, are key elements of this story, and this interplay provides abundant scope for contentious knowledge politics. But there is another level of politics at play: because of Kenya's position, what is on one level clearly a national policy process, on another level has important regional, continental – and commercial – implications (and has therefore attracted attention and intervention from a much broader range of actors). In all of this it is easy to forget the agronomist's traditional plant, plot and farm-level concerns.

Problem framing is an important focus for knowledge politics, and it is well recognised that simplistic framings can lead to years of fruitless debate. In his exploration of local seed economies in South Sudan and Tanzania, Westengen (Ch. 9) demonstrates the limitations of the framing that contrasts 'formal' with 'informal' seed systems. The supposed characteristics of these different kinds of seed systems are poorly reflected in the case studies, suggesting a disjuncture between policy and advocacy discourse on the one hand, and the more complex and nuanced reality of smallholder farmers on the other. These framings are underpinned by different ideologies, and the resulting politics must be recognised and analysed if a more productive framing and associated research agenda are to be developed.

Development-oriented agronomy is now pursued in a context that is preoccupied, or perhaps obsessed, with accountability, scaling and impact. As a result, the line between research and extension is nearly impossible to discern. It is right that this mash-up should also be reflected in *Agronomy for Development*. Leeuwis *et al.* (Ch. 5) use their experience within the CGIAR's Humidtropics[5] programme to reflect on the uncomfortable encounter between big ideas like scaling, social inclusion and stakeholder platforms; ambiguity about systems; and less than optimal institutional, management and funding arrangements. The irony that this analysis highlights is that the place-based, systemic, long-term and learning-based approaches required for scaling and impact are the very ones that the CGIAR struggles to embrace. Shifting from a large systems-based research programme to a small social enterprise, Moseley (Ch. 6) suggests that the new mantra of impact at scale combined with business principles and market models, results in a reversion to old and largely discredited models of technology transfer. In a development vision dominated by notions of agricultural revolution through introduction of technology and engagement with value chains, there is little room for participation, site specificity or local knowledge. Indeed, in this context Glover et al.'s view of technology as only coming alive through local processes of unpacking, adaptation and hybridisation is an anathema. Continuing with the theme of new models for promoting the technology and technical packages from agricultural research, Rwamigisa et al. (Ch. 7) provide an analysis of the politics of extension reform in Uganda. With significant international support, and widely heralded as the future of extension, the attempt to refashion the provision of agricultural information along private sector lines was both contentious and ultimately unsuccessful. This chapter shows how policy processes like these can be systematically analysed, with the notion of policy coalitions being placed at centre stage. Regardless of the strengths or weaknesses of the big idea that underpinned the reform, this analysis demonstrates the critical importance of everyday institutional politics. This kind of analysis can shed important light on many other initiatives and processes of interest to development-oriented agronomy – from CAADP[6] and AGRA[7] to the CGIAR reform.

The current focus on scaling, and agronomic research achieving impact at scale, can mask another critically important dimension of scale in development-oriented agronomy. Dynamics across scales – from the global to the very local – figure to some degree in all of the chapters referred to above. However, they are the centre of attention in Amanor's analysis (Ch. 3) of decades of effort to promote irrigated rice production on the Accra Plains. The forces bearing down on the irrigation schemes he describes are initially geo-political, and subsequently reflect structural shifts in the Chinese and Brazilian agricultural sectors as they embrace the Northern agribusiness model. The argument is that recent involvement by China and Brazil (in Ghana's rice sector and more broadly in Africa) has been framed as South–South cooperation, while on the ground it takes the form of technology transfer and market development for commercial advantage. Within this, consideration of locally generated or adapted agronomic knowledge is pushed aside by investment deals and resulting struggles over resources. As with some of the other examples explored in this book, an ideological commitment to market-based approaches and public–private partnership appears to undermine other potential development and technological pathways.

Most of those involved in agricultural development, including agricultural researchers, recognise that change produces both winners and losers. However, if differential impacts are dealt with at all it is usually through an economic lens as an analysis of trade-offs or benefits and costs. Fraser (Ch. 10) argues that as agronomy has become more oriented to development it has been opened up to normative contestation around rights and justice. Specifically, development-oriented agronomy is increasingly being assessed in normative (i.e. are its effects 'just', 'fair', 'virtuous,' 'right,' or 'good'?), as well as factual terms (i.e. did productivity or nutritional status increase?). The potential for contentious politics around normative judgements such as these is boundless, but, helpfully, Fraser proposes a framework which he suggests provides a basis for integrating considerations of rights and justice into both agronomic research and analysis of the politics surrounding it.

In the final chapter (Ch. 11) Giller *et al.* reflect on some of the framings that hobble the ability of development-oriented agronomy to address the challenges of sustainable agriculture and food, and how the promise of a new golden age for development-oriented agronomy might be realised. The danger is that we blink and the opportunity is lost.

Acknowledgment

Research for this paper was partially funded by CGIAR Research Programs on MAIZE and WHEAT.

Notes

1   https://dl.sciencesocieties.org/publications/aj/about
2   https://library.cgiar.org/bitstream/handle/10947/3746/CGIAR%20Strategy%20and%20Results%20Framework%202016%E2%80%932025%20-%20Final%20Consultation.pdf?sequence=1
3   The conference was sponsored by the STEPs Centre, the Futures Agricultures consortium, the International Maize and Wheat Improvement Center (CIMMYT) and Wageningen University.
4   See   https://www.worldfoodprize.org/en/laureates/2016__andrade_mwanga_low_and_bouis/ (accessed 30 October 2016).
5   https://humidtropics.cgiar.org
6   The Comprehensive Africa Agriculture Development Programme – http://www.nepad.org/programme/comprehensive-africa-agriculture-development-programme-caadp (accessed 14 February 2017).
7   The Alliance for a Green Revolution in Africa – https://www.agra.org/ (accessed 30 October 2016).

2

ON THE MOVEMENT OF AGRICULTURAL TECHNOLOGIES

Packaging, unpacking and situated reconfiguration

Dominic Glover, Jean-Philippe Venot and Harro Maat

Introduction

In this chapter, we examine how farming technologies move between places and how they are unpacked and 'grounded' in particular spaces and contexts. We argue that a better understanding of how this process occurs helps to shed light on a source of contestation within agronomy. We discuss two farming technologies that have been at the centre of controversial debates among experts, policy makers and the wider public: the System of Rice Intensification (SRI) and drip irrigation. We argue that these technologies have been contested partly because important social dimensions have been neglected, which have led to the technologies being configured and appreciated differently in different sites. Here, we use the term sites to include farmers' fields, experimental stations and laboratories, research and training centres, as well as discursive spaces such as agricultural and natural resource policies and research publications.

We selected the cases of SRI and drip irrigation in order to show that different technologies can shed similar light on the socio-technical configuration and discursive politics of farming technology. Drip irrigation epitomises a modernist package of engineer-designed hardware procured from outside the local farming systems, while SRI is promoted as an agro-ecological, low-external input methodology that relies chiefly on locally available natural resources, labour and farmers' skills. However, both technologies are associated with discourses about natural resource conservation and increased farm productivity. In practice, both have been transformed into technology packages in order to help make them mobile, and both are necessarily subject to processes of unpacking when they arrive in particular locations.

We develop our argument as follows. In the next section we discuss different understandings of technology and its role in agricultural development. We challenge the notions of 'technology transfer' and 'scaling up', which still hold key places in agricultural development narratives. We favour an alternative conception of how technologies move from place to place, emphasising the reconfiguration of relationships among individuals, social groups and material resources, and the transformation of practices in a particular time and place. This agent-centred, practice-focused understanding of technology leads us to recognise 'technology transfer' as an attempt to reorder farmers' practices by introducing new objects and instructions. We then turn to the cases of SRI and drip irrigation in order to illustrate our practice-based notion of technology and technological change as expressions of situated socio-technical practice. In the final section, we argue that our analysis provides an insight into the contestations surrounding these two particular technologies.

Understanding technology and change in agriculture

Agronomists and agricultural engineers typically conceptualise farming technologies as assemblies of consumables and equipment, or as packages of technical and managerial practices. Conceived in this way, the introduction of farming technologies to new settings, or their movement from one rural site to another, is seen as a matter of distributing artefacts accompanied by instructions and training. Technological change is seen as a simple, merely technical process, epitomised in conventional accounts of how innovations 'diffuse' and how technologies 'transfer' (e.g. Ruttan and Hayami 1973; Rogers 2003; cf. Glover *et al.* 2016).

This technicist conception of technology frames technical objects and their associated instructions as manifestations of objective scientific knowledge. As such, technologies are thought to have fixed functional characteristics that produce predictable effects. In such accounts, non-technical and non-economic factors are considered externalities, and they are often blamed when the results of technology transfer fall short of expectations. Low levels of uptake or disappointing impacts are often attributed to factors such as an unfavourable institutional framework, a lack of leadership or political will, insufficient financial investments, and even the ignorance and backwardness of uncooperative farmers. These obstacles are often targeted for correction through training and 'capacity building', while the design, delivery or performance of the technical intervention itself may go unquestioned (Glover 2010).

Since the 1980s, this conception of technology transfer and adoption has been strongly attacked, especially the privileged place it gives to scientific forms of knowledge and practice. In particular, scholars highlighted how traditional notions of technology transfer and adoption largely ignored farmers' agency and capabilities. This oversight effectively excluded local knowledge, which was often vital for producing lasting and positive change in farming outcomes. Social scientists pointed out that farmers might have good reasons to deviate from the prescriptions of agronomists and consequently it was important to understand their perspectives, values and priorities (e.g. Loevinsohn and Kaiser 1982; Chambers and Jiggins 1987a, 1987b; Chambers *et al.* 1989).

However, the challenge was neither merely to transfer scientific knowledge into farmers' practice, nor to celebrate farmers' innate wisdom, but to promote dialogue between different systems of knowledge and practice (Richards 1985; Thompson and Scoones 1994). Farmers began to be recognised not merely as important end users of technology packages but as key actors in 'agricultural innovation systems'. Innovations were seen to emerge from multiple sources through the interaction of different knowledge sets, including those of farmers as well as professional scientists and engineers (Biggs 1990; Biggs and Clay 1981; Douthwaite *et al.* 2001).

It became clear that the deployment of similar technical artefacts among different actors in diverse contexts would give rise to many different, site-specific technological configurations. Farming itself could be conceived as a combination of technologies enacted through practice, situated in time and space and within a specific social and agro-ecological context (Richards 1989, 1993). Through this situated enactment of technology, locally specific configurations of social and technical components emerge. Technology itself could be understood as a hybrid of social and technical components, comprised not only of technical artefacts and practices but also the agency of human actors embedded in a web of social and ecological relationships, expressed in many different cultural and institutional forms (Richards and Diemer 1996).

A key reason why technologies change as they travel is that they are not encapsulated in artefacts (tools, instruments, gadgets) nor even in abstract knowledge, but enacted through situated practice (Shapin 1998; Dowd-Uribe *et al.* 2014). This means that a technology is necessarily embedded in particular social structures, symbolic practices and material conditions, and that the technology will be altered when those structures, practices and conditions are different, or modified (Pfaffenberger 1992).

Typically, new farming technologies are developed in well-resourced environments such as agricultural research stations, where cultivation is supervised by teams of technicians and often involves heavy fertiliser and chemical use. Even when experiments are conducted under farm conditions, scientists typically assemble and coordinate an array of social and material resources in a particular time and place, in order to bring unruly field conditions under control and observation. This effort helps to make the 'field' more like the 'laboratory' and allows the scientists to collect the kind of scientific-but-realistic data they need (Henke 2000; Maat and Glover 2012). The technologies that emerge from this situated practice are indelibly marked by it.

From this perspective, the processes involved in the initial development of a new technology and those involved in putting it to use may be recognised as distinct (albeit linked) enactments in situated practice. In these distinct enactments, different, site-specific configurations or hybrids of technical, social and institutional components are created and re-created. In other words, the 'same' technology will in fact be different when, and because, it is embedded in different sites, and enacted by specific networks of people and groups interacting with local material resources and biophysical conditions. Instead of asking how a technology can move from the laboratory or research station into farmers' fields, a new question comes into focus: how can the skills and 'placeless' knowledge of professional scientists be made relevant to, translated into and integrated with forms of knowledge and practice that make sense on the ground?

To help a technology leave the experimental setting and travel to a farm situation, it must first be made mobile. It must be detached and made independent from the specific circumstances where it was developed, conveyed to a different location, and unpacked in that new context. The process by which this happens is the main focus of our discussion. To develop our analysis, we use the concepts of *inscription* and *affordance*, as developed by anthropologists of technology and Science and Technology Studies (STS) scholars.

Inscription and affordance

Inscription describes the work done by technology designers and engineers to build into new technical objects their expectations about how those objects are to be used. Inscriptions are programmes that call on the user to adopt certain modes of action and behaviour in order to use the object and achieve the outcomes intended by the designer. Inscription can thus be interpreted as a way of disciplining users to fall into line with the expectations and intentions of other social actors – often more powerful actors, who are trying to achieve their own objectives. This is highly visible in factory assembly lines, where a series of machines and work-stations are assembled and configured in a particular spatial order, so that trained operators may perform a sequence of choreographed steps in a complex manufacturing process. To carry out their tasks properly, the operators are required to conform to the expectations of the machines' designers, engineers and owners. Here, the inscription of a particular mode of interaction into technical objects is complemented by more obviously social and cognitive disciplines, imposed for example by training, incentives and punishments, team working, collective responsibility and managerial supervision. But the disciplining of factory workers' activities is also an intrinsic part of the function of the technical objects themselves (Latour 1991; Callon 1991; Webster 1991; Akrich 1992; Murdoch 1997).

Outside factories, everyday objects also embody the expectations of their designers and seek to govern the behaviour of users. In the case of the large and unwieldy fobs attached to hotel keys, disciplining the users – not to take the key when leaving the hotel – is an essential part of the object's purpose, deployed by the hotel manager in an effort to impose his will on his guests even when he is not present in person. In this manner, objects extend their designers' programme of action through time and across space and serve to solidify social relations including hierarchical relationships of dominance and subservience or resistance. Technical artefacts and technology packages embody scripts or programmes of action that users can adopt, subvert, change, resist or ignore (Latour 1991; Callon 1991; Webster 1991; Akrich 1992; Murdoch 1997).

A basic part of the designer's task is to consider how his or her designs are intended to be used, for what purposes, and by whom; this includes anticipating the capacities and proclivities of the proposed users, as well as potential deviations from the plan that might be undesirable (from a given point of view) or perhaps dangerous. This means that designers and users are caught up in a sort of dialogue, because the users also have work to do in interpreting what a technical object is intended for and how it may be used, whether for its intended or alternative purposes. The power of users to (re-)interpret or re-purpose technical objects – the power of 'de-scription' in the difficult jargon of STS scholars (Akrich 1992) – sets a practical limit on the extent to which designers and engineers can effectively discipline users. Users can and commonly do also resist, subvert or bypass, in both trivial and significant ways, the modes of action expected and inscribed by designers (Suchman 1987; Woolgar 1991; Latour 1991; Akrich 1992; Latour 1992). This theoretical insight from STS provides a conceptual language to make sense of forms of resistance to and subversion of technological interventions that have been documented by scholars in development studies, history and anthropology (e.g. Loevinsohn and Kaiser 1982; Scott 1985; Van Damme *et al*. 2014; Maat 2015).

To examine in detail how technical objects mediate in the relationship between designers and users, the concept of *affordance* is helpful. Affordances are the potential options for use to which technical objects lend themselves, or the opportunities for interaction that are built into the objects' designs. Affordances are partly aspects of the materiality of objects – their physical properties and characteristics, which enable and constrain the ways in which those artefacts may be employed. There are also situational or relational aspects, which is to say that the affordances may depend on the context and the capabilities of the potential users to recognise and act upon the potential uses. This perspective gives equal priority to the materiality of the objects in question and their interpretive flexibility in the perceptions of different social actors. Together these define the objects' capacities for uses and applications of various kinds, including functional, ritual and symbolic uses that may not have been intended or anticipated by the designers (Pfaffenberger 1992; Hutchby 2001).

Making technology mobile: Packaging and unpacking

To make agricultural technologies mobile, ready for transmission to farmers' fields, work is done to refine and standardise them into packages. This is an exercise in inscription. The recommended package of practices is not an exact copy of what has been developed and tested in an experimental setting, but a distillation and selection of practices, techniques and inputs deemed by technical and communications experts to be correct, coherent and appropriate. This intellectual and practical process involves simplifying the technology into a manageably small number of essential components, which are supposed to be widely applicable. These are inscribed in the form of guidelines, recommendations, schedules, checklists, equipment and kits. Through these inscriptions, the technology is detached conceptually and materially from the particular place and context where it was developed so that it may travel to new settings. While it exists in this form the technology package may change further as it passes from one organisation to another, upon translation into a new language, and/or through the design of new training modules and materials.

The concept of affordance is key to understanding what happens when the technology package arrives in each new setting. The package is not simply 'adopted', that is, received and put to use. From the potential user's perspective we can think of the technology (inscription) at this stage as a kind of proposition comprised of a set of ideas and material components, in other words, an offer or invitation to which individuals, groups or communities have an opportunity to respond.

The affordances of the technology package help to determine what may be done with it, or how potential users may respond to the proposition. A new socio-technical configuration might emerge from the encounter between the inscribed technology package and new actors and contexts, but if so it will not be a simple case of technology transfer but a site-specific, *sui generis* enactment. Because small-scale agriculture often involves collective action and coordination within households and even across communities, the enactment of farming technology in a new situation implies a redistribution of responsibilities and activities among individuals and groups (McFeat 1972, Hutchins 1996; Richards *et al.* 2008). As the introduced technical knowledge and techniques are interpreted, evaluated and integrated with existing local social and technical resources, local social and cultural structures and systems (such as those governing the coordination of labour) are also modified (Pfaffenberger 1992; Glover *et al.* 2016).

From this perspective, it is axiomatic that a technology is not the same wherever it travels; it will change wherever it 'touches the ground'. A similar package of tools and methods is likely to be deployed in a more or less different way in each place by different sets of actors. In other words, the material facts of rice physiology or hydraulics may be singular but those that are salient in one situation are liable to be different from those that matter in another.

The notion that technology is enacted through situated practice is a perspective that is difficult to reconcile with the placeless discourse of science, in which technologies represent applications of scientific knowledge that are robust and general, not particular to the actions or beliefs of individuals or groups. This discourse is often used by programme designers, policy makers, journalists and marketing professionals when they evoke the potential of agricultural technology to transform the livelihood opportunities of poor people, whereas the idea of enactment frames technologies as expressions of the capacity of people themselves, including poor people, to use tools, apply skills and organise themselves in order to solve problems and achieve goals.

In the sections that follow, we illustrate these arguments using the cases of SRI and drip irrigation. We first provide some background and briefly describe the actor-networks in which these two agricultural technologies have emerged. We then describe how they were transformed into packages to make them mobile, the attempts to inscribe certain socio-technical relations, and the processes through which users have reinterpreted and reconfigured the technologies beyond the expectations of the designers. We draw from our own fieldwork and interviews as well as from our collaboration and interactions with the Masters and PhD students who conducted in-depth field work in India, Morocco and Burkina.

The System of Rice Intensification (SRI)

SRI is a method of rice cultivation that has attracted considerable international attention over the past decade. The system involves a combination of low-external input cultivation methods for raising seedlings, establishing and nurturing them in the main field, including minimal irrigation, wide spacing, soil aeration, mechanical weed suppression and fertilisation, preferably using organic sources. The SRI method is capable of producing good rice grain yields while economising on seed and water (Uphoff 1999; Stoop *et al.* 2002; Stoop 2011; Kassam *et al.* 2011; Uphoff *et al.* 2011; Berkhout *et al.* 2015; Gathorne-Hardy *et al.* 2016).

SRI has been promoted with much energy and enthusiasm by an international network of scientists, non-governmental and civil society organisations (NGOs and CSOs), farmers' groups and others. SRI promoters emphasise that the system is a productive and ecologically sustainable method that is accessible to resource-poor cultivators because it does not require costly external inputs, such as improved seeds or agricultural chemicals. SRI advocates also argue that the method is intrinsically adaptable because it is based on a set of flexible principles rather than imposing fixed practices. They frame SRI as the opposite of Green Revolution-style crop intensification, which is often characterised as the adoption of standardised technology packages centred on high-yielding crop varieties in association with chemical fertilisers and irrigation (Uphoff 1999, 2002, 2007). For example, an important principle of SRI is that rice seedlings should be given ample space to access soil nutrients, oxygen and water, but the precise planting density to be adopted in a given situation is supposed to be adapted to suit the characteristics of the rice variety being planted, the richness of the soil in important nutrients, the risk of waterlogging, and the length of the growing season. Similar guidelines, with local adaptations, are supposed to apply to the desirable number of seedlings to be planted in each hole or 'hill', the irrigation schedule, and so on (Glover 2011a).

SRI has been the subject of heated contestation among scientists. When it first attracted scientific attention, some agronomists and economists disputed some of the scientific claims made by SRI advocates, and questioned the originality of others (Surridge 2004). One dimension of the dispute has been institutional rather than purely scientific. SRI was not developed on an agricultural research station but compiled from various sources by a field-level agronomist, development worker and Roman Catholic missionary, Father Henri de Laulanié, who worked in relative isolation in Madagascar. The inductive, empirical and 'bottom-up' origins of SRI are often emphasised by its supporters as a key attribute (Glover 2011a; Stoop and van Walsum 2013; Chavez-Tafur 2013). Meanwhile, SRI has been questioned by some influential scientists associated with prestigious university departments and research institutes, while grassroots supporters and field-level promoters of the system have reported outstanding results (Glover 2014).

One axis of the dispute has been that critics of SRI have disbelieved reports of success emerging from the field. They say that the cultivation methods actually used in some reported cases of success have appeared not to conform to the strict definition of SRI best practice. SRI's supporters respond that the critics have failed to grasp that flexibility is an essential feature of the system. On the other hand, when some mainstream scientists have tested SRI cultivation methods against alternatives, and come up with negative findings, they themselves have been attacked by supporters of SRI for failing to apply the SRI model properly (Dobermann 2004; McDonald *et al.* 2006, 2008; Uphoff *et al.* 2008). This has led some of the critics to complain that SRI advocates seem to claim credit on behalf of SRI whenever the results happen to be positive, regardless of whether the cultivation practices actually used conform to precepts of SRI that are thought to be essential and which are supposed to distinguish it from conventional methods.[1] They also complain that SRI is untestable if its definition is so flexible and vague. Our argument is that this contestation can be explained, at least partly, as a consequence of misconceptions that surround the nature of technology as a situated practice or enactment, and the way technologies are packaged and unpacked as they move from one place to another.

Making SRI mobile: simplification and standardisation

Henri de Laulanié explicitly recognised that his recommendations concerning rice cultivation had been designed in and for the particular context of a poor, highland rice-growing community in Madagascar during a period of economic difficulty. His knowledge of rice cultivation was integrated with his understanding of the local agricultural system, including off-season crops and livestock. Although he was confident that his recommendations for rice cultivation were firmly rooted in certain essential features of rice physiology and morphology, he nonetheless made clear that his methods should be understood as general principles that needed to be adapted for use by other communities in different agro-ecological settings (Glover 2011a).

In order to make SRI mobile, de Laulanié began the process of inscribing its basic principles, and this was taken further by others after his death. Through this inscription, SRI was simplified and standardised so that it could be communicated to others. It was translated into a much less flexible package, typically summarised in the form of a list of specific technical practices and parameters (see Box 2.1). These guidelines have helped to carry SRI well beyond its origins in recent years, however, in the process, many of the nuances and details discussed by de Laulanié at length have been downplayed. For instance, in SRI extension guides, the desirable spacing distance between seedlings is very commonly specified as 25 cm, while considerably less emphasis is placed on the desirability of adjusting the planting density to suit the rice variety, season, or local soil conditions. Other parameters are typically defined with similar exactness, downplaying the desirability of making local adjustments. Correspondingly, little attention has been given to developing heuristics that could help farmers or extension workers to make adjustments to suit particular locations (Glover 2014).

BOX 2.1 INSCRIPTION OF SRI

Henri de Laulanié's discussion of the techniques making up SRI was extensive, detailed, and nuanced (de Laulanié 1993, 2003). Today, the SRI method is typically summarised concisely as a short checklist of technical practices, often specified rather precisely, along the following lines: (1) raising seedlings in a thinly sown, well-fertilised, irrigated and carefully weeded nursery; (2) uprooting and transplanting seedlings when they are very young (ideally 8–12 days and not more than 15 days old); (3) transplanting single seedlings, widely spaced, in grid patterns (typically this is specified as 25×25 cm); (4) sparse irrigation to promote moist, aerated soil conditions, ideally including dry periods of 3–6 days (this is sometimes known as alternate-wetting-and-drying irrigation, AWD); (5) a regular weeding schedule, typically four times at 10-day intervals after transplanting and ideally carried out with a mechanical rotary weeder that churns and aerates the soil, hand-weeding without aeration being the second-best option; (6) fertilisation using organic sources (manure, compost, green manure crops) to the extent possible. This list slightly exaggerates the flexibility allowed in many real-world cases of SRI extension, since it has been compiled from several different peer-reviewed articles as well as field research observations in Madagascar, Nepal and India. In practice, the SRI methods have often been conveyed to farmers as a remarkably inflexible package of practices (Uphoff 1999; Stoop *et al.* 2002; Berkhout and Glover 2011).

Enacting SRI in new places

The story of how SRI touches the ground in specific sites in India is a story of farmers' and communities' intricate navigation of a locally specific range of social, cultural, institutional, agro-ecological and other factors. Typically, SRI has been introduced into communities by an external agency of some kind, usually an NGO or CSO, or a government extension agency or programme. As such, SRI usually arrives in the form of information, often conveyed through orchestrated events such as training courses and demonstrations, as well as visual presentations and printed manuals. In many instances, SRI information is accompanied by inducements such as financial or in-kind subsidies, often including mechanical implements such as rotary weeders (for weed suppression and soil aeration) and line markers, roller markers or planting frames (for marking out muddy fields with a regular planting grid or rows). It falls to the farming community and individuals within it, as well as extension staff, to work out what will be done with this package of information, artefacts and financial resources.

Multiple studies confirm that the full suite of SRI methods is quite rarely implemented. The typical pattern is that individual components of the SRI approach are used or adapted quite selectively, and practised alongside existing techniques. For example, seedlings are rarely transplanted singly when very young, because the operation is more demanding with tiny young seedlings and brings greater risk of seedling mortality, which obliges farmers to refill resulting gaps later on. Sparse irrigation or AWD can only be practised where farmers have good control over their water supply and drainage, or where there is effective cooperation among groups of neighbours in an irrigation command area. Where labour is scarce, flooding is an effective and labour-saving way of suppressing weeds. Testimony of farmers confirms that their practical decision-making about which methods to apply in each of their rice plots takes into account multiple considerations, including the characteristics of the field (soil quality, drainage characteristics, topography), distance from the home (close fields can be supervised more intensively), availability of sufficient labour at key times, access to water, land tenure arrangements, and other factors (Glover 2011b; Ly *et al.* 2012; Noltze *et al.* 2012; Berkhout *et al.* 2015; Sen 2015).

SRI is often portrayed in academic papers as an open-ended script that affords a high level of flexibility for farmers to rewrite and localise it. The modularity of SRI's components is supposed to facilitate this flexibility because the individual components may in principle be taken up independently. In actual practice, SRI is often promoted rather inflexibly as a fixed package. Moreover, the tailoring of SRI to fit a local situation is not really a question of selecting among the six practices, but giving particular specifications to the resulting practices, such as an appropriate seedling age for transplanting a particular rice variety in a given season, and so on. Heuristics to assist farmers to make these local adaptations are often lacking (Glover 2014). Local adaptations nevertheless emerge, which demonstrates that SRI's affordances are created partly by the situations and the agency of the farmers who unpack SRI and reconfigure their rice farming system, and not only by the designers' inscription.

Drip irrigation

Drip irrigation is a system of irrigation whereby water is applied to each plant in small, frequent and precise doses through a network of perforated pipes and emitters. Drip irrigation originated from the drawing tables and laboratories of engineers and researchers. The technology emerged in the 1960s in Europe, Israel and the United States on experimental research stations managed by irrigation engineers (Venot *et al.* 2014). Experiments showed that drip irrigation could save water and labour and increase crop productivity compared to conventional irrigation systems. Proponents also argue that drip irrigation enables the extension of cultivation into areas that could not be irrigated previously.

Drip irrigation systems reflect a modernist, engineering approach to irrigation, geared primarily towards large-scale, intensive farmers in developed economies. Some development actors criticise the effort to target drip irrigation towards smallholder farmers in developing countries. High capital costs and intensive management requirements are seen as obstacles for small-scale farmers to benefit from the technology. Over the last decade, NGOs, social enterprises and industrial manufacturers have tried to re-design drip irrigation systems to fit the realities of small-scale cultivators in low-income countries. This has mostly involved making systems that are smaller, easier to use, and cheaper (Postel *et al.* 2001; Venot 2016).

Proponents of drip irrigation agree that the equipment has the potential to distribute scarce water and meet crop irrigation requirements much more efficiently (Doorenbos and Pruitt 1977; Doorenbos and Kassam 1979). Contestation arises largely around technical and material questions, such as the best technical design and whether the equipment is suitable for smallholders (e.g. van der Kooij *et al.* 2013; Venot *et al.* 2014). On a larger scale, contestation revolves around the actual water saving that widespread uptake of drip irrigation would entail compared to alternative irrigation systems (Seckler 1996; Perry 2007).

Making drip irrigation mobile: extreme inscription through kits

Standard textbooks typically promote an idealised model of drip irrigation comprised of uniform rectangular fields and neat drip irrigation lines. The standard layout for capital-intensive farms (Figure 2.1, left panel) hardly differs from the design for smallholder systems (Figure 2.1, right panel). Both include the same general components: a water source, a filtering system, a network of evenly spaced, neatly aligned pipes, and geometrically regular blocks representing fields or field sections. Diagrams such as these are key parts of the inscription of drip irrigation, which help to make the technology mobile (Box 2.2). The inscription also includes features not visible in these drawings, for example the need to maintain a uniform flow and pressure within the pipes. How this is to be achieved in particular cases is left to farmers themselves or to field-level irrigation system designers.

Enacting drip irrigation in new places

Similarly to SRI, the story of how drip irrigation is enacted is highly site-specific. Morocco and Burkina Faso, for example, provide contrasting pictures. In Burkina Faso and in other countries of sub-Saharan Africa, drip irrigation is usually introduced into communities by an NGO or a government extension agency. The equipment typically arrives on the back of a pick-up truck, in the form of artefacts packed into cardboard boxes accompanied by instructions that are conveyed to farmers through community meetings, training courses and demonstrations, visual

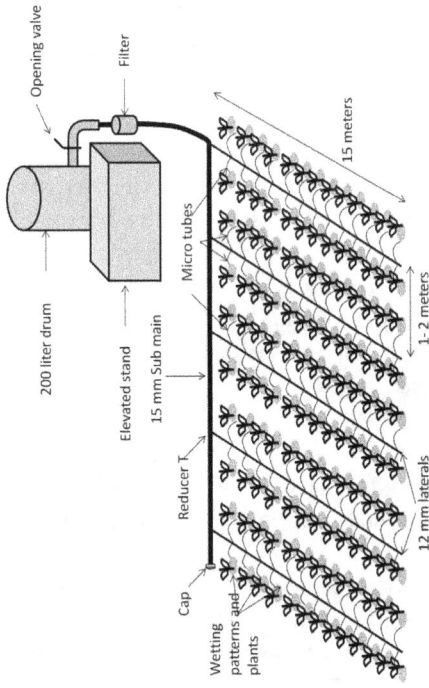

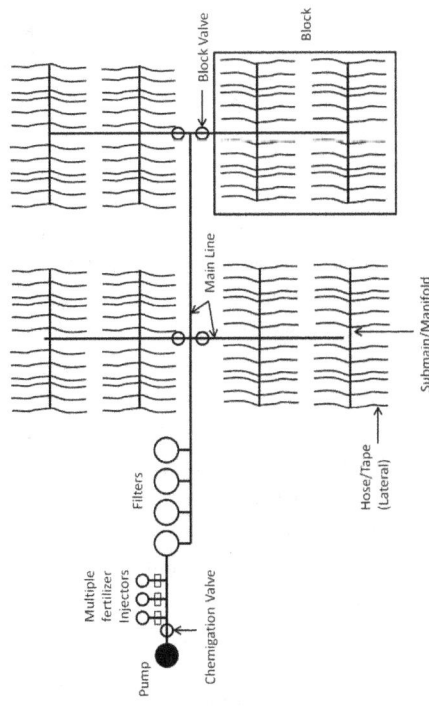

FIGURE 2.1 Simplified drip irrigation system layouts (Sources: left panel – Burt and Styles 2007; right panel – Polak and Yoder 2006)

BOX 2.2 INSCRIPTION OF SMALLHOLDER DRIP IRRIGATION

The inscription for drip irrigation typically specifies the following features: (1) the system components – water tank, control valve, filter, mainline, laterals, micro-tubes; (2) the shape of the field served by the system – a rectangle of specified length and width; (3) distances between lateral pipes and between emitters; (4) the height and volume of the water reservoir; (5) the required flow; and (6) the schedule of watering.

A separate aspect of inscription, particularly evident in developed and transition economies (such as Morocco) occurs through the establishment of norms and standards, which are technical specifications developed by engineers and tested in experimental laboratories. These standards typically relate to indicators such as the relation between water flow and pressure, line resistance, head-losses, the uniformity in irrigation, and the emitter's resistance to clogging.

The most striking instances of inscription are 'drip kits', which are complete packages of drip irrigation equipment. Drip kits are conceived as modular units that can be infinitely combined one to another; they nonetheless come in a selection of standard field sizes. Twenty square metres, 100m², 200m², 500 m² and 1,000 m² are the most common, usually with a predetermined rectangular geometry of specified length and width. Although the technical artefacts distributed in drip kits are theoretically usable by any farmer, in the minds of their designers and promoters kits of different sizes are destined for use by different classes of farmers. Smaller kits are meant to be used by women, medium-sized kits by men and larger ones by cooperating groups of farmers.

Promoters of drip irrigation systems for smallholders in developing countries invoke the need for iterative 'user-based design' rather than a top-down approach of 'technology transfer'. By listening to farmers, the engineers can inscribe their feedback into redesigned and adjusted equipment (Polak 2008). Among the most noticeable adjustments observed are alternative sizes of systems and the types of emitters used (built-in emitters or 'micro-tubes').

The explicit targeting of drip kits towards different kinds of farmers reflects the engineers' gendered view of smallholder farming in developing countries, but it also suggests a key way in which the theoretically scale neutral and flexible nature of a modular kit system may be undermined by assumptions that are built into extension strategies. Likewise, the affordances of a drip kit may be modulated by gendered assumptions about how the artefacts should be used, and by whom. In other words, the affordances of the technology package are not wholly determined by the physical characteristics of the artefacts included in it, but also by the expectations of promoters and potential users.

presentations and printed manuals. In many instances, the information and artefacts are accompanied by inducements such as seeds, fertilisers and pumps, not to mention the positive status attached to being a pilot farmer in a development project (Wanvoeke *et al.* 2015, 2016).

In Morocco, drip irrigation reaches small-scale farms in a different way. Many smallholder farmers first encounter drip irrigation while working as labourers or managers on large farms owned and managed by Spanish entrepreneurs, who have imported drip irrigation equipment from Spain. As well as acquiring skills in operating drip irrigation systems from these farms, smallholders often reuse equipment that has been discarded by the large operators. Early smallholder drip irrigation thus emerged from tinkering or *bricolage*, in which individual off-the-shelf pieces of irrigation equipment were modified and adjusted (Benouniche *et al.* 2014). A local network of small-scale manufacturers and retailers emerged, specialised in producing and marketing relatively low-tech ancillary devices (such as filters) and reconditioned second-hand equipment. Informal networks of farmers, agricultural merchants and self-proclaimed experts exchange knowledge and provide practical answers to specific problems (Poncet *et al.* 2010; Benouniche *et al.* 2011, 2014).

In Morocco farmers are advised to bury the pipes in order to protect the plastic tubes from sunlight and accidental damage. However, it is common to see smallholder drip systems where the lines are installed above ground rather than buried (Figure 2.2). The pipes used in these above-ground systems are usually smaller and lighter than the ones in below-ground installations. This adaptation is useful because many farmers rent their land and change plots from year to year. Also, having the tubes on the soil surface makes it easier to trace and repair blockages and leaks.

Many farmers have also inserted small valves on each of the lateral lines instead of a single large valve on the secondary pipe (Figure 2.2). This allows them to direct water to parts of their field as needed instead of watering the whole area at once, which helps them to cope with uneven terrain and to manage mixtures of crops in the same field that have different growing seasons.

FIGURE 2.2 Widespread adaptation of drip systems by Moroccan farmers (credit Maya Benouniche)

In Burkina Faso, inscription by engineers is stronger and, consequently, drip irrigation is less common among smallholders. Those few small growers who do use drip equipment have reconfigured the technical artefacts in various ways. For instance, it is very common to observe smallholders wetting the soil each morning, using hoses or buckets, then filling the water reservoir of the drip system before heading off to other tasks. This happens notably in the early stage of plant growth and is often repeated in the evenings, when farmers return from their day of work. Another common adaptation is the insertion of a greater number of micro-tubes than recommended by designers. Both modifications have been made to secure sufficient water supply to crops. The efficiency of these adaptations might be questioned by agronomists, nonetheless they constitute a locally specific enactment of drip irrigation, using the equipment in ways that were not envisaged by engineers and water managers.

The cases of drip irrigation in Morocco and Burkina Faso illustrate the locally contextual character of a technology package's affordances. In Morocco there was a good fit between local farming systems, which had long been oriented towards fruit and vegetable production, and the dry farming systems for which drip irrigation was originally designed. Second, patterns of employment and informal networks created channels for learning and sharing knowledge and skills between farmers. Finally, the modular nature of the hardware and the availability of second-hand equipment and consumables allowed small cultivators to acquire the artefacts they needed at low cost, and to adapt them to suit their own circumstances and capacities. Through their enactment of alternative types of drip irrigation, farmers detached the technology from its origins in large-scale commercial farming and reconfigured its social and technical components to suit their own needs.

In Burkina Faso, drip irrigation is not yet commonly used in the fields of small-scale farmers because drip kits are inscribed with expectations and assumptions based on contexts with a history of irrigation, relatively easy access to water resources through rural electrification, and a supportive informal sector. Drip irrigation has also been promoted as a new type of farming, which is expected to replace existing farming systems and enhance livelihood outcomes. In other words, the affordance of drip kits in Burkina Faso is restrictive rather than enabling, offering limited opportunities for engineers' and farmers' systems of knowledge and practice to hybridise.

Discussion and conclusions

Our analysis of SRI and drip irrigation suggests that agronomy is contested when different actors have different perspectives on the functions and purposes of technology, leading them to espouse conflicting views about how the technology should be implemented and used. This leads to disputes around definitions, standards and specifications, 'proper' versus 'improper' implementation, how the technology should be evaluated, and so on. In reality, technology does not come in the form of neatly transferable packages but is enacted by specific actor-networks in particular contexts. This means that we should expect technology to be different in different places and circumstances. Disputes arise partly because those involved have inappropriately rigid expectations based on idealised models and norms.

SRI and drip irrigation are both socio-technical systems that were developed in particular settings. They were both made mobile through processes of inscription that transformed them into packages. These packages arrive in new settings as propositions, to which local individuals and communities may respond (or not). In order to be put to work, the technology packages of SRI and drip irrigation need to be unpacked, literally and metaphorically, and (re)configured by the potential users.

The SRI and drip irrigation packages are inscribed with programmes of action, which embody the expectations and assumptions of their designers and developers about how good farming should be done and by what sorts of farmers. These scripts represent an effort to influence farmers' behaviour, regardless of whether that effort is motivated by public or private motives, benign intentions, indifference to farmers' perspectives and priorities while emphasising other goals (e.g. 'feeding the world'), or even hostility (e.g. a conviction that ignorant or incompetent farmers are obstacles to technological progress and sustainability). The inscription may incorporate, in addition to a technical prescription, ethical norms (e.g. sustainability, agro-ecology, or efficiency), modes of commercial engagement (e.g. production of a marketable surplus) and policy narratives (e.g. promoting economic growth or food security).

SRI and drip irrigation can both be understood as hybrids in a double sense – between social and technical components; and between the socio-technical worlds of agronomists and engineers on one hand and farmers on the other. A successful configuration of newly introduced technology occurs when the perspectives of designers hybridise with those of farmers. This happens more easily if the affordances of the proposed technology facilitate ready incorporation of new artefacts and practices into farmers' repertoires. Because the affordances feature in the technology package not only by inscription but also through their relation to the local context, the potential for reconfiguration and the ease with which it can be accomplished also depend on the existing capabilities of individuals and communities, as well as the modes of extension used. There is some evidence that participatory extension methods such as farmer field schools, and the existence of a healthy 'skilling' dynamic in the community (Stone 2011; Stone 2016) make it easier for farmers to learn about and experiment with the new information and artefacts they encounter. This area deserves further research attention.

This theoretical argument has several practical implications. First, making the inscriptions explicit will allow the designers and managers of technological interventions to reflect on their assumptions, better identify the presumed domain of any technological package, and consider their mission, goals, and targeting strategy. Second, maximising the affordances of technologies, in a way that is sensitive to the capabilities of target groups, calls for them to be designed and presented as flexible and malleable, in order to encourage farmers to select and adapt them to suit their circumstances. Such an approach would reduce the motivation to create rigid guidelines for the sake of replicability and scaling up. Third, the relational, contextual character of affordances suggests that the role and utility of 'pilot farmers' as a key focus for interventions should be reassessed. The farmers selected to pilot new technologies are usually chosen because they are regarded as 'leading' or 'progressive' farmers, whom others will follow. However, those other farmers may well have distinct needs, goals and circumstances which mean that they need to incorporate and adapt the new technology in a different way rather than trying to emulate the leading farmer. Therefore, instead of seeing to it that pilot farmers embody an idealised concept of the 'good farmer', and that they adopt and demonstrate the technology 'properly', extension efforts should assist farmers and their neighbours to consider and evaluate the new proposition, so that they may decide whether and how to incorporate it into their existing systems of knowledge and practice, reconfiguring these in the process.

Acknowledgements

Our analysis is based on research conducted by two distinct research groups at Wageningen University and their respective partners. The Knowledge, Technology and Innovation group coordinated the research project entitled 'The System of Rice Intensification as a social-economic and technical movement in India' from 2010 to 2014. The Water Resources Management group coordinated the research project entitled 'Drip Irrigation Realities in Perspective' from 2011 to 2015. Both projects were funded by the Dutch Organisation for Scientific Research (NWO). We acknowledge the work of our project collaborators who carried out fieldwork on SRI and drip irrigation that has informed our understandings and arguments which we present in this chapter, particularly Debashish Sen and Sabarmatee on SRI and Maya Benouniche and Jonas Wanvoeke on drip irrigation. Finally, we wish to thank participants in the panel '"Grounding": when multiple ontologies meet material facts' at the conference of the European Association of Social Anthropologists (EASA) in Tallinn, Estonia, 31 July–1 August 2014, and the second Contested Agronomy conference at IDS, Brighton, 23–25 February 2016.

Note

1   This complaint was aired by some critics of SRI in relation to a report of record-breaking rice yields in Bihar, India in 2012 (Diwakar *et al.* 2012). As well as disputing the record yield claim itself, it was noted that the excellent results were attributed to SRI even though the farmer concerned had used modern rice varieties and other external inputs.

3

SOUTH–SOUTH COOPERATION AND AGRIBUSINESS CONTESTATIONS IN IRRIGATED RICE

China and Brazil in Ghana

Kojo Amanor

Introduction

China and Brazil have framed their increasingly visible role in agricultural development in Africa as South–South cooperation (SSC). The defining characteristics of the SSC framework are third world solidarity based on mutual interest, non–interference in domestic affairs, and horizontal learning and exchange between the development partners (Alden *et al.* 2010; Golub 2013; Amanor and Chichava 2016; Scoones *et al.* 2016). This is often contrasted with the northern development framework that is portrayed as being based on the imposition of conditionalities, institutional reforms promoting normative values, and unequal relations emanating from a history of colonial and neocolonial domination. SSC draws upon the construction of a moral framework of diplomatic relations between nation states originating in the Non–Aligned Movement and the New International Economic Order and integrates this with technology transfer. Thus SSC contests the dominant modes of technology transfer within development and the paternalism that underlies conceptions of economic aid. However both China and Brazil have emerged in recent years as agribusiness powers with highly developed fertiliser, machinery, seed and biotechnology sectors, which influence and shape their agrarian policies. Northern transnational corporations have also invested in agriculture and agribusiness in China and Brazil and heavily influenced their subsequent development (Abdenur and da Fonseca 2013; Kragelund 2015; Mawdsley 2015; Amanor and Chichava 2016).

This chapter uses the restructuring of agriculture within China and Brazil as a backdrop against which to explore contestation associated with their agricultural development cooperation within Africa. To examine the various forms of contestations that occur under the framework of SSC the chapter focuses on irrigated rice production in south-east Ghana. This includes three irrigation projects – Ashiaman, Dahwenya, and Afife/Weta – and the surrounding areas. These projects originated with the construction of dams by the Soviet Union in the 1960s. In the 1970s they came within the orbit of Chinese aid, and in the 2000s were caught up in market liberalisation initiatives to encourage private sector development. The chapter examines various attempts to promote modern rice development in these areas, attempts which reflect a multipolar world, where production is dominated by private sector investment from various sources, including southern ventures (McEwan and Mawdsley 2012; Golub 2013; Mawdsley 2015). From a critical perspective on SSC, the chapter deals with contestation around the relationships between technology, innovation, and access to land, inputs and markets.

South–South cooperation: agribusiness and technology transfer

SSC focuses on diplomatic relations between states, geopolitics, and the structure of the world order and unequal trade. It also encompasses technology transfer. The UN Working Group on Technical Cooperation among Development Countries, which was established in 1972, sought to create networks for mutual learning and technology exchange among developing nations to counter increasing domination of their economies by multinational corporations (Golub 2013). The notion of technical cooperation between southern countries formed an important part of North–South dialogue of the 1970s, and later was formulated as SSC. North–South dialogue was fiercely contested by successive US governments, which instead promoted neoliberal theories and reforms in order to open up national markets to multinational corporations, sweep away state protection of national industries, and privatise state enterprises. While North–South dialogue was marginalised with the rise of neoliberal policies and structural adjustment, the closely related notion of SSC gained ground within regional and international forums, and came to fruition as the new rising powers, including China, Brazil and India, developed new trade interests in Africa and elsewhere (Hoffman 1982; Taylor 1998; Golub 2013).

During the structural adjustment era both Brazil and China competed for contracts to construct infrastructure including roads and dams in Africa. Brazil's interest in Africa dates back to the oil crisis of the 1970s, when it looked to Nigeria and Angola for oil, and new markets for its manufacturing and agri-processing industries. China has a much longer history of contact with Africa dating back to the early independence period, when it attempted to build an anti-imperialist front based on third world solidarity, and offered aid to African states to help them build self-sufficient economies (Ogunsanwo 1974; Chau 2014). Agricultural projects were a significant component of this aid.

During the late 1970s and early 1980s both Brazil and China went through an internal restructuring of their agricultural sectors. Foreign investment in agricultural input supply was encouraged, and the state supported domestic agribusiness enterprises to compete internationally. In China in the late 1970s co-operative farms were disbanded and the Household Responsibility System was introduced, based on private smallholder production (Zhang and Donaldson 2008; Li *et al.* 2009). This was associated with the rise of seed corporations, a focus on hybrid seed, closer integration of research with input production, and stronger links – through various forms of contract farming – between households, input providers and food processors. This shift essentially introduced the structures of international agribusiness relations emanating from the USA into China.

The structure of Brazilian production began to change during the 1990s with the rise of the agricultural input and food processing industries (Jank *et al.* 2001). The expansion of agricultural mechanisation and the increasing use of fertiliser and agrochemicals provided new incentives for growth. Bolstered by government initiatives, tractor sales expanded rapidly during the 1990s, reaching 40,000 units per year by mid-decade (de Monteiro Jales *et al.* 2006). Within Brazil the promotion of social protection including cash transfers to the poor facilitated the integration of family farming with agribusiness: through state subsidies, poor households had greater access to inputs and farm machinery (Patriota and Pierri 2014).

The agribusiness sector expanded rapidly in both countries, leading to increasing takeovers and concentration and the emergence of some of the largest global agribusiness corporations, including the Brazilian livestock processor JBL and the fruit juice company Citrovita. In 2016 ChemChina, the world's largest fertiliser and farm machinery company, initiated a takeover of Syngenta.[1]

The imposition of structural adjustment on African countries opened up their economies to external investment, from which the rising powers also benefited. Chinese corporations for example invested in many privatised state agri-processing enterprises in the 1980s, which lay the foundations for subsequent investments in African agriculture (Anshan 2008). A major motivation within China and Brazil was to build markets for machinery, seeds, chemicals and other products. Chinese initiatives currently revolve around agricultural demonstration sites, which promote Chinese machinery, inputs and seeds (El-Namaky and Demont 2013; Xu *et al.* 2016). They also provide training courses for African agricultural development staff to gain familiarity with Chinese technology (Tugendhat and Alemu 2016). Similarly, the main Brazilian initiatives have focused on promoting access to Brazilian machinery and technology through the More Food International Programme (Cabral *et al.* 2016); access to Brazilian cotton and rice varieties and technologies in Sahelian countries; and joint research programmes between African and Brazilian research institutions (Patriota and Pierri 2014; Amanor and Chichava 2016). Although the flagship Brazilian agricultural development programme in Mozambique, ProSAVANA, has components that encourage commercial Brazilian and Japanese large-scale farming ventures, these work within a broader framework that promotes uptake of agribusiness technologies by commercial and smallholder farmers (Fingerman 2015; Shankland and Gonçalves 2016). These initiatives all involve promoting the interests of a nexus of state and private companies manufacturing and providing agricultural technology and inputs. Critically, the choice of specific technologies and firms is influenced by complex political economy factors, rather than technical deliberations between governments.

SSC is based on an assumption that states and national policy serve the interests of their citizens rather than those of the powerful. However, developments in global agribusiness since the 1980s have resulted in increasing private sector participation – through input producers and food processors – in agricultural research and development (Buttel 1991). The world's Big Six agricultural input companies currently control 75 per cent of global agricultural R&D.[2] Although Chinese and Brazilian companies compete with northern agribusiness in Africa, they also cooperate in the development of new commercial technologies adapted to African conditions. China works with the FAO on seed research in Africa, and Chinese research institutes and corporations are working with Monsanto and the Bill & Melinda Gates Foundation (BMGF) to develop 'super hybrid' rice varieties (Gates 2013). Similarly, Embrapa, the Brazilian agricultural development agency, implements its agricultural development assistance projects in Africa in collaboration with Japan, USAID, BMGF, DFID and other international agencies. The model used in the ProSAVANA project in Mozambique is derived from the experiences of the Cerrado region in Brazil in which Japanese technical cooperation helped to develop commercial production, and aims to replicate this and facilitate the uptake of both Brazilian and Japanese technologies by Mozambican farmers (Fingerman 2015).

The main forms of contestation within and around SSC relate to diplomatic relations between states, trade policies, and conditionalities that protect markets for northern multinationals. Thus, Brazil took the US to the World Trade Organization (WTO) for unfair subsidies on cotton in 2004. The WTO panel found in favour of Brazil and the US agreed to pay compensation (Woodward 2009; Patriota and Pierri 2014). Brazil celebrated its victory by establishing the Cotton-4 programme, among four West African countries (Benin, Burkina Faso, Chad and Mali) that had also protested against US cotton policy. However, the Cotton-4 programme essentially provided these countries with access to Brazilian cotton varieties and technologies. It did not create a framework to challenge the structure of the global cotton market for the benefit of poor producers.[3]

Following a brief reflection on agricultural policy, the next section traces the development of irrigated rice production in south-east Ghana from the 1960s to the present.

SSC and irrigated rice in Ghana: history, rhetoric and reality

Agricultural policy

During the mid-1970s Ghanaian agricultural policy began to highlight a greater role for agribusiness. This shift drew upon the World Bank's model for smallholder contract farming based on nucleus estates and outgrowers, and from Ivorian experiences with this model (Daddieh 1994). In Ghana the model was developed for oil palm, irrigated rice and vegetable production. It involved the development of seed production capabilities, seed and input markets, and contractual relations which gave parastatal corporations control over production and food processing (Konings 1986; Gyasi 1992; Daddieh 1994; Amanor 1999).

In the late 1970s crop breeding programmes began to bear fruit resulting in the release of certified open-pollinated cereal varieties by national research institutions. However, the adoption of structural adjustment disrupted these developments, since conditionalities involved privatisation of agricultural services, reduction of state involvement in agricultural input distribution and removal of subsidies. Attempts to privatise the seed sector were reluctantly introduced by the state, but there was little enthusiasm among private investors to take on enterprises with questionable commercial viability. As a consequence private seed companies failed to materialise, and an association of private seed growers with little commercial acumen found itself with the responsibility for providing commercial seed (Amanor 2010). The result has been underfunded research, a lack of new varieties and seed quality issues. In recent years a coalition of donors, foundations and NGOs worked with large transnational corporations to facilitate the uptake of imported hybrid seeds and create commercial seed markets (Amanor 2010). The most significant initiatives include the Alliance for a Green Revolution in Africa's (AGRA) support for private seed companies; and a coalition involving USAID, Dupont (which took over Pioneer Seeds) and the US-based NGO, ACDI/Voca. The Ghanaian state also plays an important role in these coalitions and although input markets have been privatised, many of the networks that developed during the era of state input distribution continue to exist. During the 1970s much of the procurement of subsidised inputs for the development of large-scale agriculture in northern Ghana was conducted for the state by Henry Wientjes, a Dutch national settled in Ghana. In 1979 Wientjes registered his commercial interests as Wienco, which identifies itself as a Dutch-Ghanaian private company. Wienco now dominates the supply of agricultural inputs in Ghana, and works closely with the state to import and distribute inputs for cocoa and food crops. It also has close links with Yara (a leading fertiliser distributor in Africa), RGM (the main distributor of Syngenta products in West Africa) and many farmers' associations through which it distributes its products. Alongside these programmes has been vigorous civil society opposition to the Plant Breeders Bill, which has been portrayed by Food Sovereignty Ghana as paving the way for the introduction of GMO varieties.

Irrigated rice in southeast Ghana

Commercial rice production in Ghana emerged from the country's close relations with China in the 1960s, relations that were part of the struggle to build a world front against imperialism. In October 1962 the two countries signed an economic protocol, which included the development of irrigated rice cultivation, and during his visit in 1964 Zhou Enlai advised the government to promote rice cultivation. Fifty Chinese experts were subsequently sent to Ghana to develop the agricultural and textile sectors (*Peking Review* 1962; Ogunsanwo 1974; Chau 2014; Djansi 2015); however, their work was cut short by a coup d'état in 1966, following which Ghana's new military government expelled 430 Chinese technical experts.

The expelled experts were eventually replaced by agriculturalists from Taiwan, who developed irrigated rice projects at Dawhenya, Ashiaman and Afife. They extended the irrigated works originally constructed by the Russians and introduced new cultivation techniques for rice and vegetables (Amanor 2015). However, Taiwanese involvement in the rice sector ended following another coup d'état in 1972. The National Redemption Council re-established diplomatic relations with China and broke off relations with Taiwan: between 1972 and 1976 Chinese experts were again managing the irrigation projects.

According to farmers and retired project staff the technologies introduced by the Chinese and Taiwanese experts were similar. P. K. Anamang, a staff member at the Ashiaman Irrigation Project, recalled that the returning Chinese initially consulted closely with the Ghanaian staff, questioning them in detail about the activities of the Taiwanese. They continued to implement similar technologies and used a similar management style.[4] In her study of Taiwanese and Chinese irrigation projects in Senegal, Baker (1985: 405) makes a similar point:

> Despite their differing ideologies the teaching methods used by Taiwanese and the Chinese differed very little, both having been derived from long-established methods used in pre-revolutionary China, but also featuring many modern developments.

Critically the Ashiaman and Afife projects were designed around gravity-fed irrigation systems. This proved to be a boon in the 1990s, when other irrigation schemes collapsed as they were forced to pay commercial electricity rates for water pumping. Both the Chinese and the Taiwanese introduced an assortment of technology including specialised tractors, tillers and seeds. Irrigated plots were given to farmers to work on an individual basis and the technicians instructed farmers on cultivation techniques. Some new land management practices were introduced including the use of mulch and fertilisers combinations. Some farmers argued that when the Chinese left they were unable to continue using some new techniques such as nursery bed preparation and stubble ploughing because they lacked power tillers and specialised tractors. As a result they reverted to broadcasting seed and burning rice stubble. However, a significant number of the farmers believe that the training given by the Chinese has continued to influence their farming strategies and has re-enforced their own experimentation (Amanor 2015). For instance, Simon Gbododzor stated that:

> Those of us who worked on the Chinese farm in those days were not many. The Chinese showed us many things about rice cultivation. They trained us so well that after they finished their contract and left we were able to teach

the other farmers. We were the ones who really transferred the knowledge they brought.[5]

By the late 1970s China's commitments to agricultural development projects in Ghana waned as it began to restructure its aid policies towards Africa. Chinese foreign policies shifted from an emphasis on winning diplomatic support through economic assistance to one of supporting capital accumulation through foreign investment. The Chinese agricultural projects in Ghana were returned back to government and they were put at the centre of a variant of the World Bank smallholder contract model. However, as a result of the severe economic crisis of the late 1970s, which brought on the collapse of state-sponsored commercial agricultural production (Shepherd and Onumah 1997), this attempt to develop contract smallholder production was not successful. By this point most of Ghana's irrigation facilities were seriously dilapidated. Under structural adjustment donors were unwilling to fund state-sponsored irrigation development. Moreover, assistance from the US government prioritised export crops and open market policies that enabled US food products to penetrate international markets (Bello 2004). As a result of Ghana's acceptance of these policies, US rice came to dominate the market before being overtaken by perfumed varieties from Thailand and Vietnam.

During the 1990s support for irrigation projects came primarily from Japan and the EU. In accord with dominant donor policies of liberalisation and removal of government subsidies, the emphasis was on introducing community management and user fees to encourage cost recovery. Although new techniques and varieties were introduced to farmers, they were provided at market cost, which many farmers could ill afford. At Dawhenya the project collapsed as farmers struggled to pay user fees. The Electricity Corporation of Ghana cut off power supplies because bills remained unpaid. A few farmers continued to work on their irrigated land by using their own private generators (Kranjac-Berisavljevic et al. 2003).

The main irrigation projects that continued to operate during the 1990s were based on gravity flow, such as those constructed by the Chinese at Afife and Ashiaman, and they became the main areas of rice production. The Ghana Irrigation Development Authority (GIDA) sought to maintain control over smallholder rice production in these areas through contractual relations with farmers: GIDA arranged loans for seed and other inputs, and extension advice, and the loans were repayable in kind after harvest. Contracts were also arranged with marketing companies, the Agricultural Development Bank (ADB), and input suppliers. Inputs were purchased in bulk from Wienco with assistance from the ADB. However, these programmes were marred by high rates of default by farmers, delays in input distribution, poor quality seed, inappropriate fertilisers, and problems with water control (Kranjac-Berisavljevic et al. 2003; Obirih-Opareh 2008). Most of the farmers did not accept the recommendations of GIDA and continued to farm autonomously and to use their own rice varieties. Although the contract farming schemes collapsed, the rice produced by farmers was popular on the market and competitively priced, and market traders from Kumasi travelled to Afife to purchase rice and offer farmers loans (Amanor 2015). However, the farmers had very slim profit margins, and were beset by problems in gaining access to reasonably priced credit and quality, cost-effective inputs.

In recent years rice production has come back as a cornerstone of national agricultural development policy, with the main focus being to encourage investment by foreign capital in large-scale production. The focus is now on development of rice estates, or nucleus estates with outgrower schemes in which farmers are provided with seeds, inputs and credit. These different arrangements involve various types of linkages between private corporations from different nations, the state and civil society organisations. Corporations from Brazil, China, the US and elsewhere are now competing to gain a foothold in rice production.

Private foreign investment

The continued production of rice on irrigated areas in the Accra Plains convinced the government that despite the problems of the past, Ghana had a comparative advantage in commercial rice production. Beginning in the mid-1990s, increasingly frustrated in its attempts to gain donor backing for irrigated rice development, the government sought to attract medium-scale foreign investment to the rice sector. The first arrangement was with the US-based Quality Grains Company, with the aim of developing rice production on a 4,300 ha plot of land, originally acquired by the government for cotton production. This ended in disaster: the CEO, Juliet Cotton, squandered the US$ 20 million the Government of Ghana invested in the project and was imprisoned in the US, while the Minister of Finance, the Minister of Food and Agriculture (MOFA), and other prominent officials were imprisoned in Ghana.[6] Subsequently, a new venture was formed with Prairie Volta from Texas, in which the government and Ghana Commercial Bank each held a 30 per cent share. However, the project was delayed for eight years as surrounding communities challenged Prairie Volta's right to the land. Production only began in 2009 on 750 hectares. Difficulties in gaining financial support, resulting from the reluctance of banks to release funds while the land issue remained unresolved, resulted in further delays (Anderson 2011; Amanor 2015).

Another significant private investment was made by Brazil Agro Business Group, which started operating in 2008. It acquired a lease for 5,000 ha at Kpenu in the Volta region, involving land which had originally been leased to a Norwegian biofuels company. With 500 ha under irrigation the company acquired rice seed from Prairie Volta and multiplied it at Kpenu. It also drew upon technical networks within Rio Grande do Sul in Southern Brazil to introduce innovations such as cheap but effective means of constructing irrigation ditches, and techniques for germinating rice under water, which enable three crops to be grown per year. However, the company is constrained by the technology – including seeds – available in Ghana, and the high costs of inputs and machinery. Brazil Agro Business is also embroiled in disputes with the surrounding communities over land, with rival factions in different communities contesting ownership and the conditions under which the company gained access to it (Amanor 2015). Although Brazil Agro Business has brought Brazilian technicians and innovations into Ghana, it does not have formal support from either Embrapa or any Brazilian banks.

Brazil Agro Business inspired another company, Global Agri-Development Cooperation (GADCO), to established rice cultivation in the Volta Region. GADCO is a medium size company, which is registered in Amsterdam; its founding members originate from Nigeria, Britain and India. Its Chairman is Baron Malloch-Brown, a former UK government minister in the Foreign and Commonwealth Office, United Nations Deputy Secretary-General and Administrator of the United Nation Development Programme. With funding from the German Development Bank (KfW), GADCO combines production on a nucleus irrigated estate with an outgrower scheme. Unlike Brazil Agro Business, GADCO has not acquired large tracts of land but has entered into an innovative arrangement with the community to lease 1,000 ha in return for a 2.5 per cent share of gross earnings and a promise to release 48 ha of irrigated landback to the community (Osei 2012; Amanor 2015). GADCO makes use of many of the technologies introduced by Brazil Agro Business. It contracted Agropecuária Foletto, a company involved in rice production in Rio Grande do Sul with links to Brazil Agro Business, to provide management services and introduce Brazilian rice technology. However, Agropecuária Foletto experienced difficulty in importing the technology into Ghana, and could not find local equivalents, so GADCO has now entered into an alliance with Wienco. This new arrangement gives GADCO potential access to Syngenta seeds and inputs, and some influence on agricultural policy in Ghana. GADCO and Wienco are also developing outgrower linkages with farmers in the Afife/Weta Irrigation Project,[7] who are provided with seed and technology packages by Wienco, and loans by GADCO, in exchange for selling their rice to the company at harvest. GADCO has released its own Copa Connect brand of rice on the Ghanaian market (Amanor 2015). Agropecuária Foletto has moved out of Ghana.[8]

A final example is ChinaGeo, one of the two biggest Chinese construction companies active in Ghana, which has moved into commercial agriculture. The company developed a small vegetable farm at Aveyime on land of under 50 ha to feed its workforce, but is interested in irrigated rice production. Although Chinese interests have not been directly involved in rice production since the 1970s, in the 1980s and 1990s Chinese construction companies were active in the rehabilitation and extension of the Afife irrigation project. In 2013 ChinaGeo and the Ningxia Province Administration in China expressed interest in a programme to expand and develop rice irrigation in the Afife/Weta Irrigation Project. Feasibility studies were complicated by the World Bank's focus on developing commercial irrigation facilities on the Accra Plains using a Public Private Partnership (PPP) model in conjunction with a protocol that requires community consultations before investment deals can be concluded. Thus the Chinese feasibility studies had to be carried out in consultation with the surrounding communities. Several sections of the community expressed dissatisfaction with the proposed project, which reflected pre-existing fractures within the community including disputes between the chiefs of Weta and Afife over historical rights to ownership of the land on which the irrigation project is situated. Although the rights of the Weta chiefs to the land has recently been recognised, overturning earlier recognition of the rights of the Afife chiefs, most of the farmers on the irrigation project are from Afife. The Chinese technical evaluation team (led by MOFA) only consulted with the Weta Chiefs, as a consequence of which the Afife chief mobilised the local youth against the project. Relatively wealthy sugarcane farmers also opposed the project. They were worried that their land would be appropriated and redistributed to smallholders. Those opposing the project organised effectively: some youth physically threatened the Chinese evaluation team, which led to the feasibility studies being abandoned (Amanor 2015). However, it is likely that another factor influencing the Chinese withdrawal was the movement of GADCO and Wienco into the project site. While the technical team from Ningxia Province Administration returned to China, ChinaGeo still remains interested in rice production. However it has moved out of the Accra Plains, where the World Bank's involvement and the PPP framework reduced its room to manoeuvre. The company has shifted its focus north to the transition zone, negotiating access to land at Prang near Sunyani and in Akumadan, in northern Ashanti, where it intends to rehabilitate and extend old state irrigation facilities and develop contract farming schemes with smallholder farmers. In these areas it is negotiating for significantly larger plots than the 1,000–5,000 ha sites typical of the private sector rice investments found on the Accra Plains: the Prang land is 36,000 ha. ChinaGeo is also involved in a long-term project to build a Chinese Rice Research Institute at the Ashiaman Irrigation Project site. This will create conditions under which Chinese hybrid rice can begin to be produced and adapted to local environmental and market conditions (Amanor 2015).

Beyond rice: South–South cooperation and agricultural development in Ghana

Although a Brazilian company is at the forefront of irrigated commercial rice production in Ghana, rice does not feature in official Brazilian government cooperation with the country. The major Brazilian programme in Ghana is More Food Africa, a multi-country programme that aims to address the needs of smallholder farmers by introducing Brazilian technologies that have worked to raise living standards of the rural poor in Brazil. The first phase of the programme is concerned with importing Brazilian farm machinery and tractors for smallholders in several African countries (Cabral *et al.* 2016). It is important to note that while the technologies that are being disseminated emanate out of rural development experiences in Brazil, they are received into a different context in Ghana, where the tractors are allotted to the Agricultural Engineering Services Directorate (AESD). Unlike the Brazilian Ministry of Rural Development which is responsible for the More Food Programme in Brazil and its extension into the international development arena, the main objective of AESD is to facilitate the Agricultural Mechanisation Service Centre (AMSEC) Programme, *not* to address rural poverty. AMSEC promotes private commercial agricultural mechanisation centres that are supposed to build demand for tractor services among rural farmers. This programme is not at all concerned with building linkages between family farming and social protection, or adapting technology recommendations to smallholder needs. Thus, while the Brazilian Ministry of Rural Development has a vision of how its agricultural modernisation policies for smallholders can be applied to Africa through SSC articulated with agribusiness, the Ghanaian government has its own distinct vision.[9]

Beyond the framework of official programmes many Chinese companies are investing in Ghana's agricultural sector. For instance, Chinese herbicides dominate the local market: they are inexpensive and widely used by smallholders. Many Ghanaian traders are also involved in importing herbicides from China. As a result of intense competition, one Chinese company, Wynca Sunshine, has opened a factory in Kumasi to introduce further economies of scale, making use of local distributors and agrochemical dealers. Shen (2013) argues that Chinese companies are relocating to Africa as a result of increasing competition and falling profits at home, and many of these companies do not register with the local Chinese Chambers of Commerce or the Chinese embassy, preferring instead to develop their own connections into African markets (Shen 2013). Thus Chinese agribusiness also operates outside the confines of state-to-state consultations and prescriptions of SSC.

Conclusion

SSC posits an alternative model of technology transfer to the conditionalities that characterise northern aid. In principle this is based on the common experiences of third world countries and the relevance of technologies that have proven to be transformative in the rising powers. SSC is presented as an example of 'win-win', where China and Brazil gain access to new markets and African countries gain access to appropriate technology. In reality, the recent agricultural development of the rising powers has been based on facilitating northern investments within their agricultural sectors and in reorganising production along agribusiness lines inspired by the US. This includes adopting biotechnology and promoting engagement with smallholders through contract farming. A major objective of SSC is thus to gain space in which the technologies of emerging powers can compete against US dominated agribusiness, and gain access to markets. This involves attempts to gain increased representation within existing international development platforms and organisations, and articulating and building consensus for developing country interests. In contrast, in the context of technology generation these emerging nations work in close collaboration with northern agribusiness interests.

Despite the claims that Brazilian and Chinese science provide alternative agricultural technology adapted to conditions found in Africa, SSC essentially promotes mainstream northern technologies. While historically Chinese agricultural cooperation did provide alternative models which sought to promote self-sufficiency (Ogunsanswo 1974; Baker 1985; Brautigam 2009; Chau 2014; Amanor and Chichava 2016), with a new focus on hybrid seeds and inputs this has been replaced with high input agribusiness models to facilitate higher yields.

The recent history of the irrigated rice sector in south-east Ghana is littered with contestation and conflict around land, engagement with local communities, technology choice and access to quality seeds, inputs and credit. These contestations are at variance with the framework of SSC, which projects a world in which states represent the undivided interests of farmers and guide them towards rational adoption of new technologies. In reality, as seen in the case of Texas Prairie, even the state was unable to guarantee its own legitimate access to land.

The complex network of interests that intersect state, transnational capital, and civil society also acts to thwart the Chinese and Brazilian interventions into African agriculture. Thus Brazil Agro Business was able to make impressive gains in the productivity of irrigated rice, but the constraints of market regulation, and the control over available technologies by entrenched interests frustrated the intention to further extend the reach of Brazilian management and technology. Without access to Ghanaian state support, private actors from Brazil interested in rice production could not gain support from Embrapa, and without support from Embrapa they could not gain recognition from the Ghanaian state. Although the rice technologies developed by Brazilian technicians show promise, they are a product of informal networks in Ghana, rather than of Brazilian development institutions. As such they constitute a threat to the authority of Brazilian technical cooperation and the interests that it represents. This helps explain why the initial euphoria around Brazilian interventions in rice production in Ghana waned and the company involved in providing technical services to other rice production companies (Agropecuária Foletto) relocated to Kenya. Similarly, Chinese innovations in irrigated rice production introduced in the 1970s continue to influence farmers, but conflict with current prescriptions of MOFA and current Chinese agricultural policy. Thus the agricultural policy objectives of the Ghanaian and Chinese states conflict with farmers' learning and experience, and both states attempt to negotiate an agricultural cooperation that negates these historical experiences, and builds upon recent developments in agribusiness and technical advances.

Acknowledgements

Research was supported by the 'China and Brazil in African Agriculture' project (www.future-agricultures.org/research/cbaa), and the UK Economic and Social Research Council (grant: ES/J013420/1) under the Rising Powers and Interdependent Futures programme. I am also grateful for the help of Aminu Aliu and Charity Akpabey in carrying out the research.

Notes

1 See http://www.etcgroup.org/content/global-agribusiness-mergers-not-done-deal-0, 15 December 2015. Sourced on 10 January 2016 and http://www.bloomberg.com/news/articles/2015-11-12/chemchina-is-said-to-be-in-talks-to-acquire-syngenta, 12 November 2015. Sourced on 10 January 2016.

2 http://www.etcgroup.org/content/global-agribusiness-mergers-not-done-deal-0 15 December 2015. Sourced on 10 January 2016.

3 See Aid-For-Trade Case Story: Brazil. Brazilian Cooperation Agency of the Ministry of External Relations (ABC/MRE) / Project Cotton-4. 2011. https://www.oecd.org/aidfortrade/47699046.pdf (accessed 20 June 2016).

4 Interview with P.K. Anamang, Dawhenya Irrigation Project, 2 September 2014.

5 Interview with Simon Gbodzor, Afife, 2 December 2013.

6 See http://news.bbc.co.uk/1/hi/world/africa/1832416.stm, 22 February 2002 (accessed 20 November 2016); and https://www.modernghana.com/news/115198/quality-grains-scandal-bombshell-na-who-cause-am.html, 14 September 2004 (accessed 20 November 2016).

7 The Afife Irrigation Project is now known as Weta Irrigation Project following a legal case in which the Weta chiefs have been recognised as the rightful owners of the land rather than the Afife chiefs.

8 It is currently developing rice in the Tana River Region Kenya, where it has found a more receptive policy environment and where it has access to a wider range of commercial rice varieties. It is planting Bayer rice hybrids imported from India (http://www.planetaarroz.com.br/site/noticias_detalhe.php?idNoticia=12097 *Planeta Arroz* Uma semente gaúcha na África, 10 May 2013. Sourced 12 November 2015).

9 One of the objectives of the current agricultural plan is to 'collaborate with private sector to build the capacity of individuals and companies to produce or assemble appropriate machinery tools and equipment locally' (Ghana Ministry of Food and Agriculture 2007).

4

GM CROPS 'FOR AFRICA'

Contestation and knowledge politics in the Kenyan biosafety debate

Stephen Whitfield

Introduction

The Kenyan National Biosafety Authority's first annual National Biosafety Conference took place in August 2012. In her conference address, guest of honour Professor Margaret Kamar, Minister for Higher Education, Science and Technology, spoke to the audience about her first day in the Ministry explaining that it coincided with frenzied media reports that genetically modified maize was illegally entering the country. She described a situation within the Ministry of different people asking different questions, unsure whether this was an issue about the health risks of genetically modified foods, the traceability of genetically modified organisms, or the policing of Kenya's borders. In trying to make sense of this complex set of issues, she explained, 'I came to the Ministry and I asked one question… "what do the scientists in Kenya say?"'

Adapted excerpt from *Adapting to Climate Uncertainty in African Agriculture*
(Whitfield 2016: 159)

Within the context of development-oriented agronomy, it is argued by some that genetically modified (GM) crops hold the key to poverty alleviation and a productive and prosperous future agriculture (Wambugu 1999), and they have emerged as an internationally supported technology that is central to the much sought-after African Green Revolution. Through public–private partnerships and within autonomous international strategies (such as that of the Consultative Group on International Agricultural Research – CGIAR), transfers of technology that promise impact at scale are being pursued (Brooks *et al.* 2009; Sumberg and Thompson 2012). Outside of the Alliance for a Green Revolution for Africa's (AGRA) focus on seed systems, the Bill & Melinda Gates Foundation (BMGF) has partnered with Monsanto PLC's social responsibility wing and the International Maize and Wheat Improvement Centre (CIMMYT) of the CGIAR to develop and deliver GM Water Efficient and Insect Resistant Maize for Africa. Monsanto, and other giants of the global seed industry, are simultaneously investing in commercial products, such as Bt maize and cotton for the African market, and already control a large share of the seed sector in South Africa (OECD 1998). Bilateral donors, including some from European countries, have supported the New Partnership for Africa's Development (NEPAD's) Comprehensive African Agriculture Development Programme (CAADP)[1] by funding university programmes and research facilities across the continent (Okusu 2009). These bilateral partnerships as well as the work of the International Food Policy Research Institute (IFPRI) and USAID's programmes have been instrumental in developing capacities for, and guiding the development of biosafety regulation necessary for the technology to enter into African seed systems. The prospect of pipeline GM seed products in many cases has represented a necessity-based impetus for determining the rules that should govern their environmental release or traceability within food chains (Kingiri 2010), thus setting a precedent, and arguably creating a path for other technologies that may follow.

But, the development of regulatory regimes has been the subject of much contestation. Although the GM debate has been defined by passionate pro- and anti- campaigns, characterising it by this singular distinction belies the complexity of the questions that shape GM politics. In Africa in particular, the potential use of genetically modified organisms (GMOs) is about societal needs and risks, regulatory capacity, corporatisation, technological efficacy, environmental change, consumer rights, and more; and debate inevitably revolves around incomplete evidence bases and personal and political values (Herring 2007). In asking 'What do the scientists say?' Margaret Kamar's attempt to defer to a single and objective source of knowledge represents a profound misunderstanding of this complex subject of contested agronomy. It is nevertheless a misunderstanding that is both politically powerful and much replicated.

As one of the leading African nations in terms of biotechnology research, related development activities and the establishment of national regulations, Kenya is an important case. How knowledge politics around GMOs plays out in Kenya is highly significant for the future of the technology in Africa. It is also a case for which this knowledge politics has been particularly volatile over recent years, with a gradual progress towards a GM future being punctuated by moments of intense discussion and debate. This chapter aims to unpack this knowledge politics, both as a means of opening up space for an airing of those aspects of the debate that have been otherwise closed down, and to draw out broader lessons for effective governance of GMOs in Africa. The chapter reflects on Kenya's experience of developing and passing its 2009 Biosafety Act; the introduction of Labelling Regulations in 2012; and the moratorium on importation and consumption of GM foods imposed in the same year. Particular emphasis is placed on the way in which framings within the regulatory debate are path dependent, and on the role of knowledge and evidence in challenging established trajectories. This analysis is situated within a broader political economy of GM technology, in order to both contribute to its characterisation and to draw out implications for its future across the African continent.

A conceptual framework: regulation, coalitions and knowledge politics

In suggesting a broad conceptualisation of regulation, beyond the conventions of state-centric risk management, Newell (2002: 1) argues that regulatory actions may 'facilitate commercial transactions and generate public trust in new technologies', and thereby promote their public benefits. Such risks and benefits alike may be associated with technology as an end-product (e.g. a GM seed), technology as a scientific process (e.g. genetic modification), or even technology as broader change within socio-technical systems (e.g. towards a future GM agri-food system) (Jasanoff 1995). Any or all of these might be the intended subject of regulatory action in their own right, or they may be so closely interconnected that they are difficult to distinguish (Romeis *et al.* 2013). Technology regulation, then, is inclusive of actions that increase the burden of evidence about safety, uncertainty, outcomes and probabilities, as well as those that reduce it; inclusive of actions that prohibit certain uses of a technology as well as those that incentivize others; and inclusive of actions that limit the trade of technologies as well as those that facilitate it. Focusing on both broad governance and specific uses allows for a consideration of how regulation and technology development interact across the whole commodity chain, from front end innovation through to commercial production, trade and use. Although these stages might be governed by separate regulatory regimes, they are always closely interlinked (van Zwanenberg *et al.* 2008, 2011). The viability of research and development endeavours, for example, may depend on the existence of market opportunities for end products, intellectual property protection, or even financial support from the state.

Across the process of genetic engineering, and along the pathway to a future GMO-based agriculture, there are a number of potential objects and moments that might be subject to regulation. These range from the intellectual property associated with raw genetic material; through the laboratory processes of genetic mapping and gene extraction and transfer; the resultant modified seed; the in-field practices of planting, growing, and harvesting from that seed; the entry of GM foods into local, national, and international supply chains; and ultimately to the future development of the agricultural and food sector. As one might expect, as the development of technology progresses along this pathway, from fundamental science and conception to actual products, practices and supply chains, the nature of the GM debate shifts accordingly (Cartwright and Hardie 2012; Romeis *et al.* 2013). However, the reality is that neither debate nor technological development is linear: the moments, objects and objectives of a GMO pathway are conceptually and practically bound together, such that it can be difficult, both in the observation of debate and in our own rationalisation of it, to disentangle arguments about national economic and agricultural sector development from the mechanisms of traceability of GMOs within production and processing systems, and the stipulations placed on seed stability and food toxicity testing.

Existing models of biosafety regulation, and those of the European Union (EU) and the United States (US) in particular, mark the terrain for debates about regulation in Africa. The polarisation between the promotional stance in the US and the precautionary approach in the EU manifests in an absence of traceability mechanisms and safeguards in the former and a restrictive system in the latter. But rather than looking on these positions as pro- and anti-GM or as purely political abstractions, Jasanoff (1995) contrasts the 'product-based' system of the US, the focus of which is firmly on GM foods and more specifically on their metabolic profile, as the object of regulation, with the 'process-based' system of the EU, which responds to uncertainty in the scientific process and the future system and technologies to which it might be contributing (Dunlop 2000; Löfstedt and Vogel 2001; Bernauer and Meins 2003; Prakash and Kollman 2003). The different potential conceptualisations and starting points that are reflected within the broader GM debates, of which this divergence in regulatory approach is symptomatic, readily become confounded with notions of evidence, science and legitimacy. For example, in 2006 the World Trade Organization (drawing on its Sanitary Phytosanitary Measures and the Technical Barriers to Trade Agreement) judged that the European Union's *de facto* moratorium on approving new GM products, which ran from 1998 to 2004 and was based on broad concerns about genetic modification rather than specific issues with individual products, was illegal as it did not have a clear scientific basis.

The GM debate is an archetypal subject of contested agronomy exactly because of the explicit knowledge politics that plays out within it and through which trajectories of agricultural change are shaped. At a policy level, biosafety regulation, and all that is bound up within it, is a formalised manifestation of the contestations and power dynamics that take place across the different sites and scales of agricultural development. This often takes the form of closed-down knowledge and questions of knowledge legitimacy, as in the frequent 'What do the scientists say?' and 'What do they know?' arguments. From Maarten Hajer's (1995) understanding that agency within regulation debates is not just tied to individuals but also to ideas and arguments, one might conceive of knowledge as not just something that is used strategically – to advance an interest or political position – but as playing a substantive role in shaping the debate (Schön and Rein 1994: 37) by contributing to the search for, and discovery of 'intelligible solutions' (Weale 1992: 222). In order to play this substantive role, there is a degree to which knowledge must be presented as a constructed subject, one that is open to deconstruction and reconstruction through negotiation, rather than as a closed box in which inherent assumptions and value judgements are denied or unacknowledged (Stirling 1999). As is evident in the Kenyan biosafety case, the challenge of achieving inclusive governance, opening up constructed knowledge claims and avoiding the privileging of one discourse or storyline over all others is particularly acute within such a polarised political arena (Black 1998; De Marchi 2003; Jasanoff 2003; Renn 1999; Stirling 2008).

Path dependency in the recent history of GM technology development and regulation in Kenya

Genetic engineering research and the development of regulatory structures and guidelines in Kenya in the early 1990s were largely funded through the Dutch Government's Directorate General for International Cooperation. The Kenya/ Netherlands Biotechnology Programme was established in 1993 to build national capacity for research and development, including through degree programmes at the University of Nairobi; support tissue culture research and the use of molecular markers for the selection and breeding of maize within the KARI breeding programmes; and support a KARI-Monsanto partnership project (established in 1991) targeting the development of GM virus resistant sweet potato. Although part of this Biotechnology Programme involved the development of Kenya's 1998 'Regulations and Guidelines for Biosafety and Biotechnology in Kenya' within the National Council for Science and Technology (NCST – a government parastatal created within the Ministry of Education, Science and Technology, and since renamed the National Commission for Science, Technology and Innovation – NACOSTI), these regulations were of no legal consequence and effectively acted to facilitate expanding research and development activity under institutional and informal good practice guidelines (Harsh 2005). European investment in biotechnology ebbed in this early period in response to the lack of a regulatory structure and reflecting the precautionary stance on GMOs that developed in Europe in the mid-1990s (Levidow *et al.* 2000; Paarlberg 2001).

The early years of Kenya's biotechnology development, which took place largely within a 'legislative vacuum' (Wakhungu and Wafula 2004: 43), set an important precedent in the development of biosafety regulation. A lack of state structures to govern this activity resulted in those institutions involved in research and development (largely through internationally funded public–private partnerships), such as KARI, Africa Harvest, and the Centre for Biotechnology and Bioinformatics (CEBIB) of the University of Nairobi, being given a degree of regulatory autonomy. It was the ratification of the Cartagena Protocol on Biosafety in 2003 that provided much of the impetus for Kenya to develop a more formalised system of biosafety regulation. There was a shift from European-funded research and development to US investment and support to regulatory capacity building (through USAID and IFPRI). With this shift, those established biotech research institutions, as well as industry and lobby groups, such as the International Service for the Acquisition of Agri-biotech Applications (ISAAA) AfriCentre, African Agricultural Technology Foundation (AATF), and the Seed Trade Association of Kenya, found themselves in the privileged position of being the drivers of the regulatory process, both in terms of representation within the NCST and as coordinators, partners and participants in capacity building programmes. These programmes had explicit promotional agendas and aimed to provide US-based biotechnology companies with opportunities to 'transfer technology' to developing countries and thus create new markets. Over a period of ten years, and with significant capacity building support provided through the USAID and IFPRI-led Programme on Biosafety (PBS) (active in Kenya from 2004), representatives of agricultural biotechnology research and development partners participated in drafting and promoting the Biosafety Bill.

The Bill was opposed by a group of civil society organisations including Action Aid International Kenya, Africa Nature Stream, Ecoterra, Greenbelt Movement, Kenya Small Scale Farmers Forum, Kenya Organic Agriculture Network, and Participatory Ecological Land Use Management, collectively known as the Kenya GMO Concern Group (KEGCO). They mobilised around a number of campaigns and activities, which included the drafting, in partnership with a select group of MPs, of an Alternative Biotechnology and Biosafety Bill. This alternative bill was presented to parliament in 2008. The group challenged the 2004 draft Biosafety Bill on the grounds of human rights violations and the capacity and competence of national regulatory agencies as well as incomplete evidence about the health risks of GMOs. They also organised a number of public protests against the introduction of GMOs to Kenya. Although this opposition contributed to a stalling of the Bill's progress through parliament, it had little effect on its content. Following the second reading of the Biosafety Bill in Parliament in December 2008, at which it received strong support from then Minister for Agriculture, William Ruto MP, and passed with a strong majority, it was assented into law by President Kibaki in February 2009 (Karembu *et al.* 2010).

The Bill, and subsequent Act, dictates the processes and procedures for regulating GM crop testing and introduction, and outlines the roles and responsibilities of an independent body, the National Biosafety Authority (NBA), to oversee applications for GMO-related activities. It makes provisions for different aspects of GMO development and commercialisation to be regulated through specific guidelines to be appended to the Act. Interestingly, the NBA's autonomy in agenda setting has resulted in a less facilitative regulatory environment than many involved in the Biosafety Act wished for. The development and implementation of labelling regulations is a case in point, as it resulted in a challenge to the authority of the NBA by the pro-GM lobby.

In 2011, three sets of regulations addressing (1) contained use, (2) environmental release, and (3) import, export and transit, were passed into law. A fourth regulatory document concerning the labelling of GMO products for public consumption was drafted by the NBA (under the responsibilities assigned to them through the Biosafety Act) through a legal consultant, from KARI, and agreed upon in May 2012.[2] As dictated by procedure, the draft documents underwent an internal review process (through the NBA technical team) and consultation with stakeholders (invited by the NBA) across two sessions, before being submitted to the minister and the attorney general's office for gazetting.

The 2012 labelling regulations closely followed the model of the European Commission and laid out a number of legal obligations on the part of actors across the commodity chain (known as 'operators') to ensure that GM products are separated from non-GM products and traced from field to fork. This includes obligations on farmers to maintain separation distances between GM and non-GM crops to prevent cross pollination and on food companies and retailers to ensure any food, feed or ingredients containing more than 1 per cent (by weight) of a safety approved GM material has the words 'genetically modified' printed on a label or displayed at the point of sale. The stated purpose behind this regulation was to inform and protect the rights of consumers, as well as to provide a mechanism for the future recall of products if need be.

Members of the pro-GM lobby expressed dissatisfaction through the Open Forum on Agricultural Biotechnology (OFAB) which is run by the AATF and ISAAA, arguing that the cost and impracticality of the traceability requirements threaten to make GM technology unviable for smallholders. The labelling regulation designed to facilitate product recall and removal they argued would undermine confidence in safety testing and the NBA more generally (Whitfield 2016). A number of GM crop technologies, developed through charitable, donor- and state-supported programmes (such as the Monsanto Sustainable Yield Initiative and the Kenya Agricultural Research Institute – KARI) were in development in 2012. The AATF were managing a partnership between Monsanto and CIMMYT to develop drought tolerant maize and insect resistant maize (also involving Syngenta PLC), and cotton, all of which were at confined field trial stage. Trials on mosaic virus resistant cassava (involving KARI and the Danforth Centre) and viral resistant sweet potato (involving KARI, Monsanto and the Danforth Centre) had already been completed. The labelling regulations were therefore introduced in the context of an impending release of varieties that, in the case of the maize crops in particular, had been promoted and pushed as technology for Kenya's subsistence and smallholder producers. In the eyes of those investing in and developing the technology, placing a potentially costly burden on smallholders, as the potential adopters and key 'operators' in the supply chain, threatened to undermine the broader project. The NBA subsequently made some minor concessions in response to this pressure. For example, in his overview presentation of biosafety legislation in Kenya at the first NBA National Biosafety Conference (2012), contrary to the dual objectives stated in the regulations themselves, the NBA's Director of Technical Services Professor Dorington Ogoyi emphasised that 'labelling is not about safety, labelling is about consumer choice'. However, the NBA has remained relatively steadfast and the regulations themselves remain intact.

Through the Biosafety Act, and efforts to translate it into guidelines governing specific aspects of the technological development process, a facilitative regulatory regime has been emerging. But what is interesting is that the trajectory of biosafety regulation in Kenya has not been wholly predictable: just when it looked as though it was opening up to a biotech future and developing a facilitative regulatory environment this trajectory was derailed in 2012 by a decision from the Ministry of Public Health. Minister Beth Mugo MP, who had been previously peripheral to the GMO regulation debate, made the unanticipated announcement that the Kenyan Government were to implement a blanket ban on the importation and consumption of GM foods in the country with immediate effect. In her public address the Minister said that the decision had been made on the basis of 'genuine concerns that there has not been adequate research done on GMOs' and an official cabinet statement later added that 'the ban will remain in effect until there is sufficient information, data and knowledge demonstrating that GMO foods are not a danger to public health' (Owino 2012).

In contrast to Dorrington Ogoye's emphasis on consumer choice, this policy move quickly put public health back at the centre of the GM debate, and brought a new actor (e.g. the Ministry of Public Health) into a position of power within it. ISAAA and other members of the OFAB responded to the Ministry of Public Health's stated need for 'sufficient information' about GM safety. Through the International Agri-biotech and Biosafety Communication conference held in Nairobi in April 2015, the publication of global status updates and success stories, and organised tours of GM crop production sites in Europe for MPs, they put the government under pressure to reconsider the ban. In August 2015, speaking at the 4th Annual Biosafety Conference in Nairobi, now Deputy President William Ruto made the first public announcement of an intention to do so. The announcement coincided with a key moment in the development of GM crop technology in Kenya as the first applications for the release of insect resistant maize and cotton had, just weeks before, been submitted by the African Agricultural Technology Foundation. At the end of 2015, these applications were being processed by the NBA after having invited public comments.

The recent history of biosafety regulation in Kenya (summarised in Box 4.1) has been characterised by shifting coalitions and regulatory trajectories in response to key moments and emergence of new actors and evidence. The research and development of GM technology in Kenya has progressed without real restriction or limitation by the emerging regulatory structures. At the same time, the progression of research and development, and those involved in it, have actively shaped this regulation.

BOX 4.1 TIMELINE OF KEY INTERNATIONAL INFLUENCE IN THE DEVELOPMENT OF KENYA'S BIOSAFETY REGULATION

1993   Biotechnology Programme of the Netherlands Directorate General for International Cooperation begins to fund activities towards the development of a Biosafety Framework in Kenya

1996   UNEP/GEF begins demonstration project on facilitating development of National Biosafety Framework in Kenya

1998   'Regulations and Guidelines for Biosafety and Biotechnology in Kenya' introduced

Launch of phase 2 of UNEP-GEF project

1999	Establishment of Biosafety framework through UNEP-GEF support and continued support of Netherlands Directorate General for International Cooperation
2000	Kenya signs Cartagena Protocol on Biosafety
2001	African Union finalises the African Model Law (AML) on Safety in Biotechnology
2003	Kenya ratifies Cartagena Protocol on Biosafety
	Establishment of the Programme for Biosafety Systems (PBS) with funding (US$ 14.8 million) from the United States Agency for International Development (USAID) through the International Food Policy Research Institute (IFPRI)
	First full report of the Codex Alimentarius Commission *Ad Hoc* Intergovernmental Task Force on Foods derived from Biotechnology
	African Union endorses the Africa-wide Capacity Building Programme in Biosafety which encourages the adoption of the African Model Law
	NEPAD and the African Union establish the African Panel on Biotechnology
	ASARECA and ACTS begin the 'Regional Approach to Biotechnology and Biosafety Policy in Eastern and Southern Africa' (RABESA) initiative on behalf of COMESA
2006	Adoption of Kenya's 'National Biotechnology Development Policy'
	The African Union Commission proposes an African Strategy on Biosafety and declares an African Position on GMOs in Agriculture
2007	Finalisation and presentation to Parliament of the Biosafety Bill
2008	Revision of the African Model Law (renamed African Model Law on Biosafety)
2008/9	Acceptance and gazetting of the 'Biosafety Act'
2009	Replacement of the National Biosafety Committee with the National Biosafety Authority, housed within the National Council for Science and Technology
2011	Publishing of 'Environmental Release'; 'Import, Export and Transit'; and 'Contained Use' Regulations
2012	Publishing of 'Labelling' Regulations
	Ministry of Public Health and Sanitation announce a national ban on the importation and consumption of GM foods
2013	Publishing of 'Handling, storage and packaging' Regulations
2015	AATF submit environmental release application for insect resistant maize and cotton to the NBA
	Deputy President Rt Hon William Ruto addresses 4th Annual National Biosafety Conference and announces government intention to lift Ban

Unpacking the role of knowledge and power

The extent to which the various legislative processes around GMO regulation in Kenya have involved the participation (or non-participation) of different stakeholder groups has been a consistent point of argument. Harsh (2005) describes the weak and limited participation of opponents within the development of the Biosafety Bill resulting in the eventual adoption of a Biosafety Act that was largely unchanged despite a 5-year period of stakeholder consultation, parliamentary debate, and protest. The following excerpt from the minutes of a comment and response session from a workshop conducted in July 2007 by the National Council for Science and Technology (NCST) demonstrates how such workshops became an exercise in defending the framing of biosafety regulation adopted by the Bill, rather than an opportunity for participatory input. The session followed a day of presentations on the Bill by representatives from the NCST, Ministry of Science and Technology, and KARI. In the introductory remarks Dr Miriam Kinyua from Moi University (who was later appointed as chairperson of the NBA), clearly not reluctant to attempt to frame the discussion herself, 'reminded the participants that the Green Revolution passed Kenya by and there is need to look objectively at biotechnology so that we are able to reap the benefits of this technology'. The task-force panel was made up of representatives from ISAAA, CEBIB, Africa Harvest, University of Nairobi, and NCST.

> *Comment (from Africa Nature Stream)*: The Bill will be used to open the Kenyan market to GMOs.
> *Response from the Panel*: The objective of the Bill is clear: to regulate GMOs.
> *Comment (from Kenya Biodiversity Coalition)*: Small scale farmers have not been involved in the preparation of the Bill.
> *Response from the Panel*: The process began in 2002 and 15 stakeholder meetings have been held since then. Also, consistent with Kenyan laws, the Bill was published for public comment within the requisite 21 days.
> *Comment (from Kenya Organic Agriculture Network)*: The number of farmers on the National Biosafety Board should be increased.
> *Response from the Panel*: For a Board with a total membership of 16, one farmer representative is sufficient.
> *Comment (from Action Aid)*: Clause 7 does not address food and livelihood security.
> *Response from the Panel*: This is a policy issue outside the scope of the Bill.
> *Conclusion (in the meeting report)*: Following a thorough review of the stakeholder comments above, the taskforce does not see the need for any further amendments to the Bill
>
> *(OFAB 2007)*

A 21-day period for public comment on the Bill came almost four years into the drafting process and over fifteen years since genetic engineering research and the development of the Regulations and Guidelines for Biosafety and Biotechnology had begun. This restricted stakeholders' capacities to influence anything other than the finer details.

The drafting of the labelling regulations, an NBA-led process, was also criticised, but this time by the pro-GM lobby, for a similarly restricted participation process. AATF, ISAAA and others submitted comments during an open consultation period, particularly to challenge some of the technical stipulations within a draft document, such as the 1 per cent threshold of GM content, which is primarily an issue about the capabilities and accuracy of detection technology. That technical issues, as opposed to more fundamental framings and objectives, became central to consultations and debate over labelling is indicative of the limited scope of the consultation process, which again was dictated by the outcomes of an earlier and even less open drafting process.

The controversy over the GM ban and the prospect of its lifting has reopened a participatory space similar to that which characterised the drafting of the Biosafety Act. It has also opened up a parliamentary level politics in which ministries, including the Ministry of Public Health (members of which have spoken out in support of lifting the ban) and the Ministry of Industrialization and Enterprise Development (headed by biotech advocate Dr Wilson Songa) have been afforded a more prominent voice.

The prospect of the GM ban being lifted stimulated the Kenya Small Scale Farmers Forum and the Kenya Biodiversity Coalition to organise farmer-led protests, such as a demonstration march in August 2015 in William Ruto's former constituency of Eldoret. The extent to which participation in the rekindled debate over GMOs will influence the decision to lift or maintain the ban remains to be determined. Ruto's statement of intention for the government to lift the ban following a process of public and parliamentary consultation, raises questions (as there has been over previous consultations) over the purpose and scope of this consultative process and the extent to which a decision has already been made.

Although not made explicit in ministerial statements, it is thought that the ban was a response to the now infamous study published by Gilles-Eric Séralini at the University of Caen (France), in which rats fed on NK603 (Roundup tolerant) maize were more likely to develop tumours, and at an earlier stage, than a control group. The authors explained that this resulted from the metabolic consequences of the transgene in the GMO (Séralini et al. 2012: 4221). Evidence, and more commonly an absence of evidence, of environmental or public health risks associated with GM technology is commonly cited in debates about regulation, and the Séralini paper provided one of the most politically influential and contested pieces of evidence. It was also thought to be the driver behind a similar policy response in Russia (Romeis et al. 2013).

However, Séralini et al.'s work has been heavily criticised for its methodological design, criticism which the authors have countered by arguing that their methods are consistent with the toxicity study of Roundup-ready maize conducted by Monsanto (Hammond et al. 2004) and the *in vivo* tests for GMOs as outlined by the

OECD (OECD 1998). The paper was eventually retracted by the journal that published it. A statement made by the European Food Safety Authority (EFSA) in the EFSA journal explains a number of these criticisms, including the size and number of the control and test groups and an ambiguous analysis, and concludes that 'the study as reported by Séralini *et al.* is of insufficient scientific quality for safety assessments' (European Food Safety Authority 2012: 9). Although criticised, even by some within the anti-GM camp, the study has acted to call into question the scientific consensus about safety and put the spotlight on the potential for ambiguity even using the scientific method. This may have long-lasting implications for the trajectory of GM technology in Kenya, for example, by reaffirming, and arguably reframing, labelling and traceability as an issue of precaution *and* of consumer choice.

Reference to evidence as a way to legitimise some claims, and to delegitimise others (dismissing them as unscientific), has been part of the Kenyan GM debate from the beginning. Evidence bases are used both substantively, i.e. for learning and better informed decision-making, and strategically, to advance political positions. Indeed, in many ways these two objectives are difficult to disentangle. Evidence-based claims about the societal benefits of GM crops are important in justifying investment in long programme development, especially as some of these projected benefits are highly speculative. Crop yields statistics are often repeated within the policy and public sensitization campaigns by the pro-GM lobby, as evident in calls to lift the GM ban. The AATF-led WEMA project provides an example in its much-repeated claim that 'maize varieties developed under WEMA are expected to increase yields by 25 percent under moderate drought' (AATF 2010: 2). Tracing the origins of this 25 per cent figure is difficult, but it appears to be linked to experimental trials of CspB event maize (compared with its conventional hybrid) under water-limited conditions conducted by Monsanto in the American Midwest (Castiglioni *et al.* 2008). It is also linked to other studies of water-limited grain yields (Boyer and Westgate 2004; Campos *et al.* 2006), references for which were given in the WEMA application to the NBA for permission to conduct confined trials.[3] In the Castiglioni study, however, while yield averages for the GM group were significantly higher than the control groups, they find that 'the best two performing events, CspB-Zm event 1 and event 2, demonstrated yield improvements of 20.4% and 10.9%, respectively' (Castiglioni *et al.* 2008: 450), not the 25 per cent that is often cited. Water-limited studies of grain yield loss have, unsurprisingly pointed to significant negative impacts of water shortage, but the capability of WEMA varieties to mitigate against this loss can only be inferred on the basis of phenological mechanics associated with controlled conditions, of which there has been, as yet, only limited testing in an African context.

Public perception and willingness to grow, purchase or consume GM foods has become a contentious but important point of argument within the linked debate about the need for and purpose of labelling. Making reference to a USAID-funded study conducted by Simon Kimenju at the University of Nairobi (Kimenju and De

Groote 2008) an anonymous participant at the OFAB meeting in May 2012 stated that:

> You will find that 80 percent of Kenyans are actually willing to consume GM products if that is the only food they can access … therefore labelling is not really informed by the attitudes of the people.
>
> *(Anonymous respondent, OFAB, May 2012)*

The study itself asked a sample of 604 individuals in Nairobi whether they would be willing to purchase GM maize flour at the same price as their favourite brand, a question which received a 68 per cent positive response rate. This information is useful in itself, and there is a need for further and more comprehensive research into public perceptions and attitudes towards GMOs. This, of course, is also another clear example of the political nature of evidence interpretation and utilisation. The study's findings could be drawn on just as readily (and perhaps more appropriately) in making a case in favour of labelling, i.e. that a significant proportion would choose not to purchase it over their favoured brands, or that a label would not represent a severe market disadvantage.

In these cases, as is typical of the GM debate, knowledge and politics are inseparable. Evidence is used as a means to legitimise arguments, and complex motives shape the design and interpretation of studies. The assumptions, uncertainties and value judgements within evidence are often denied or overlooked, both by those generating and those utilising it, on the basis that this undermines objectivity. Doing so, however, acts to place those bodies of evidence within entrenched pro- or anti-camps – e.g. the WEMA performance tests become a piece of pro-GM evidence and the Séralini study a piece of anti-GM evidence. Acknowledging uncertainty and assumptions however, can open up space for evidence-informed negotiation, e.g. about the implications of a 68 per cent willingness to purchase among Nairobi-based respondents, or about the indicated but not yet realised potential of a GM drought tolerant variety. This entrenched knowledge politics has implications for the extent to which meaningful and productive public participation in debates can be achieved.

Discussion: fostering a progressive debate about GM regulation in Africa

Debates about biosafety regulation in Kenya have been characterised by polarisation and path dependency, but also by many twists, turns and unpredictable outcomes. This chapter has explored the power of evidence, argument and knowledge, combined with politics and power, to shift and shape these debates.

In his OFAB-invited public lecture at the University of Nairobi in July 2013, Mark Lynas argued that further debate about GMOs should be subordinate to the scientific consensus on the value and safety of this technology. Despite this call, the debate in Kenya continues to evolve. New evidence, technological developments and changing regulatory capacities will all present potential moments of change within the trajectory of agri-food system, and around which there is scope for public debate.

In some respects debate in Kenya has been shaped by those with the political power to push their agendas at the expense of others. Arguably the Biosafety Bill succeeded on the strength of agendas pushed through capacity building programmes and insulated (from alternatives) within a closed drafting process. But subsequent debate has been framed less by political agendas and more by the necessity of developing protocols for the emerging technology (as in the case of labelling) and the emergence of new information (as in the case of the ban). The need to engage with the issues of consumer choice and public health gave legitimacy to new voices, including those from outside the biotechnology industry. This has resulted in a question of the purpose of regulation (i.e. to reflect its role in protecting consumer choice) and a challenging of the idea of objective scientific evidence as the basis for legitimising perspectives.

Achieving meaningful participation and negotiation across polarised perspectives remains a significant challenge. In the drafting of the Biosafety Bill, the PBS and the NCST advanced a discourse of public participation but the extent to which this was achieved in practice was limited to an ineffectual consultation process. The same was true of the NBA-led process around the development of labelling regulations. This is in part an issue of process, and points to the need for forums that facilitate meaningful and open dialogue at the early stages, including mechanisms that ensure marginalised or non-expert voices are heard. Civil society and social research efforts have an important role to play in this regard, as do the forums and annual conferences of the NBA, which have shown potential to act as an open and productive platform.

However, even where processes permit participation, stand-offs between perspectives can persist, either because there is no bridge across which different and competing knowledges can be communicated, or because the assumptions that underpin different perspectives block exchange. Where 'contradictory certainties' (Thompson and Warburton 1985) persist, unaddressed across a process of technology development and regulation, discourse coalitions become entrenched and polarisation and path dependency are unavoidable.

We must, therefore, look not only to the process but to the substance and subjects of debate, with the aim that the debates are organised around productive questions. Debate around the nature and role of evidence is an area that could represent potentially productive grounds for the interaction of polarised perspectives. For example, can questions such as 'how much evidence is sufficient?' or 'how can we increase certainty about agronomic performance?' be used to move beyond the pro- and anti-GMO stalemate? Making space for negotiated perspectives requires a critical reflection on the framings that each brings to the debate (and potential alternatives), and the knowledge (and knowledge gaps), evidence, and values that underpin them. Without recognition of, or in the context of denial of, the uncertainties associated with a GM future – for example, the performance and capabilities of GM seeds; the consequences of cross-breeding with wild relatives; the metabolic consequences of transgenes and the expression of new allergens; the operator costs associated with traceability; and more – alternative perspectives can be readily dismissed as unscientific and the space for communication and negotiation limited. Without this critical reflection on evidence we are left with an unsatisfactory 'what do the scientists say?' type of justification of what is highly politicised evidence and argument.

In the advancement of modern biotechnology and associated regulator regimes across Africa, there are lessons to be learnt from the stalled trajectory and stand-offs, as well as the successful interactions, that are evident in the Kenyan experience. Although unlikely to prevent undesirable outcomes for some, which are probably inevitable, given the contested and complex nature of the GM debate, there is a clear need to achieve meaningful negotiation across perspectives. This will only happen when there is acceptance of and critical reflection on the uncertainties and ambiguities that transcend pro- and anti-GM perspectives.

Notes

1 CAADP Pillar 4 'aims to improve agricultural research and systems in order to disseminate appropriate new technologies': Forum for Agricultural Research in Africa (2011).
2 A fifth regulation, regarding 'handling, storage and packaging' was drafted in 2013.
3 It is notable that it was in this application that I should first come across references to crop performance studies, as biosafety applications are, according to the protocols of the Biosafety Act, predominantly about demonstrating safety, not efficacy.

5

SYSTEMS RESEARCH IN THE CGIAR AS AN ARENA OF STRUGGLE

Competing discourses on the embedding of research in development

Cees Leeuwis, Marc Schut and Laurens Klerkx

Introduction

The Consultative Group on International Agricultural Research (CGIAR) was founded in 1971, and currently operates under the banner of the 'CGIAR Consortium' with a membership of 15 Research Centres. Over its lifetime, the CGIAR has been subject to numerous organisational reforms in light of prevailing concerns such as reduced spending by governmental bodies, the wish to improve the efficiency of 'the System', and the ambition to be more effective in securing development impact (Anderson 1998; Kassam *et al.* 2004; McCalla 2014). Moreover, while some centres were established with an explicit systems focus, at various stages in the history of the CGIAR there have been calls for complementing crop- and technology-focussed research with more holistic and systems-oriented perspectives (Biggs *et al.* 2016). This chapter analyses the most recent experiences with systems research in the context of the latest reform of the CGIAR (see Kamanda 2015), which included the formation of 16 CGIAR Research Programmes (CRPs) that were meant to foster collaboration, reduce competition between the individual Centres, and above all increase development impact. At the same time, the reform is presented as a strategy to strengthen collaboration with partners, and make the Consortium more demand driven, development relevant and accountable. Thus, it resonates with a number of trends in the realm of development-oriented agricultural research that were described in *Contested Agronomy* (Sumberg and Thompson 2012).

Three of the CRPs explicitly adopted a systems approach as a strategy to operationalise the reform objectives, and support innovation in smallholder agriculture. This chapter identifies several areas of contestation in the everyday operation and enactment of one of these systems CRPs, relating in different ways to system boundaries and system dynamics. First, regarding the way in which the

CGIAR engages with systems concepts we argue that there was no agreement about the nature of the systems under consideration, or about how they may change. Second, regarding the contribution of the CRP to system change we note that there was a tension between the programme and the wider CGIAR environment in relation to how development should be conceptualised, how progress towards it should be measured, and how learning can be made to emerge from development processes. A third tension relates to the CRP as an organisational system, and revolves around how the governance of financial and human resources collided with the ambition to conduct collaborative systems research and foster demand-driven research prioritisation. The indications are that these tensions negatively affected the capacity of systems CRPs to operate as a coherent whole, and survive in the CGIAR environment.

The mechanisms and cleavages associated with these tensions are analysed in terms of competing discourses about identity and the role of place-based research (i.e. research addressing problems identified by stakeholders in specific geographical context) in securing development impact. It is argued that participants in the CRP started with conflicting assumptions and views about how research may contribute to development; the kinds of public goods that research may deliver; in which contexts CGIAR research should take place; and how (and by whom) research agendas should be set. We conclude that there is a need to rethink how the CGIAR may combine the production of 'international public goods' (IPGs) with engagement in place-based development efforts.

This chapter is based on three years of participant observation by the authors in the CRP on Integrated Systems for the Humid Tropics (henceforth referred to as Humidtropics). Our involvement began mid-2012 when we were asked to strengthen the social science and innovation studies grounding of the Humidtropics proposal. As a result, the first author operated as leader of the Strategic Research Theme (SRT) 'Scaling and Institutional Innovation' from 2013 onwards, with ample support from the co-authors and others. Thus, we participated in the yearly CRP planning workshops and management meetings, a workshop with an adjacent CRP, two workshops related to 'capacity development', and a conference on 'systems research' co-organised with other systems CRPs. Moreover, in his capacity as SRT-leader, the first author participated in meetings of the Humidtropics management team, a meeting with the CGIAR Fund Council, a workshop with all CRP leaders and the Directors of Research of the Centres, a meeting with the Board of the International Institute of Tropical Agriculture (IITA), the lead centre for Humidtropics, and a meeting with the International Advisory Board of Humidtropics. In addition, all authors were involved in supervising PhD researchers who conducted comparative, place-based research as part of the CRP, and we also held many informal discussions with researchers and managers in the CGIAR community.

Engagement with system concepts: competing or co-existing kinds of systems thinking?

'System thinking' has a long tradition in international agricultural research, both outside and within the CGIAR (Devaux *et al.* 2009; Kristjanson *et al.* 2009), and has experienced both ups and downs over the past decades (Thiele *et al.* 2001). The 2009–2015 CGIAR Research Programmes (CRP) portfolio contained three 'systems' programmes: Dryland Systems, Humidtropics and Aquatic Agricultural Systems. These CRPs were distinct from the 'commodity' CRPs where the focus was on particular crops (e.g. rice, wheat or maize). In legitimising systems research, proponents often argue that smallholders do not just grow one crop, but rather integrate a range of crops and livestock in their farming system, and that their livelihoods also depend on non-farm activities. Moreover, advocates emphasise that a holistic perspective is needed to link production issues to environmental and sustainability concerns, and consider trade-offs between different objectives. It is argued that overlooking such complexities leads to the development and promotion of inadequate solutions, and hence that understanding the contextual interdependencies in the system as well as the underlying socio-economic and institutional logic is essential for generating positive development outcomes. While this seems to be a shared starting point among systems researchers, there appears to be far less agreement on how systems should be conceptualised, what boundaries and systems levels should be taken into account, and how such systems may (be) change(d).

At a conference in March 2015 organised by the three systems CRPs (Integrated Systems Research for Sustainable Intensification in Smallholder Agriculture[1]) many different types and forms of systems research were presented. In terms of boundaries, some studies focussed on the level of fields (cropping systems) while others focussed on the farm (farming systems) or the broader ecological landscape in which a farm is embedded (landscape systems). Yet other researchers included more social dimensions in their analysis, and talked about 'livelihood systems', 'socio-ecological systems' or 'socio-technical systems'. In addition, some research was presented in terms of the networks of actors that are considered important in supporting and bringing about change, i.e. 'agricultural innovation systems'.

These different conceptions were not just a reflection of arbitrary boundary choices, but involved different ontological and epistemological ideas about the nature of systems, the way in which knowledge about them can be generated, and how system change may come about. In the general systems literature, different strands of systems thinking have been identified, and these have also been linked to agriculture (e.g. Röling and Engel 1990; Bawden 1995). Table 5.1 summarises some features of different approaches to systems thinking and their associated terminology. In the CGIAR environment there has been little open discussion about these different approaches and their implications. Researchers used very different language and terminology when discussing how societal impact may be generated through the use of systems thinking.

TABLE 5.1 Different modes of systems thinking and their associated terminology

Type of systems thinking (origin and/or literature sources)	Key metaphor and assumption depicting how systems are seen	Key change strategy implied	Associated terminology
Hard system thinking (scientific management, Taylor 1947)	Machines. Interactions in natural and social systems can be known and predicted	Engineer and optimise towards a given goal	'Scaling-out technologies that work', 'Defining impact pathways', 'Optimise the value-chain'
Functionalist systems thinking (structural functionalism, Parsons 1951)	Organisms. Systems are functional wholes, depending on relations between components and environment	Re-balance and adapt in a changing environment	'Providing baskets of "best bet" technology options', 'On-site experimentation with different options', 'Building capacity to adapt'
Soft systems thinking (Checkland 1981)	Meanings. Systems consist of people with different worldviews and boundary definitions	Foster dialogue, learning and agreement among actors	'Building capacity to innovate', 'Foster multi-stakeholder learning'
Cognitive/Autopoietic systems thinking (Maturana and Varela 1984; Luhmann 1984)	Psychic prisons. Biological and social systems tend to perceive the world through their own logic and be blind to others	Shock therapy by creating a crisis	'Putting pressure on governments', 'Lobbying and advocacy'
Political/Critical systems thinking (Jackson 1985)	Arenas of struggle. Systems are characterised by power structures that constrain system change	Coalition building, competition and negotiation	'Building coalitions for change', 'Overthrowing socio-technical regimes'
Social/Institutional systems thinking (Giddens 1984; North 1990)	Rules. Formal and informal rules are produced and reproduced in interaction, resulting in certain orders	Change rules and incentive structures	'Engage in institutional experimentation', 'Creating enabling environments'
Complex systems thinking (Prigogine and Stengers 1984)	Self-organisation. New orders emerge without central steering as the unplanned result of multiple intentional actions	Identify existing trends and opportunities arising from these	'Adaptive learning and uncertainty reduction', 'Work towards tipping points'

Source: Leeuwis and Wigboldus 2017; see also Leeuwis 2004.

One could argue that the use and co-existence of different modes of systems thinking and multiple system boundaries makes sense in view of the multi-faceted and multi-level nature of agriculture, and that this reflects positively on the CGIAR's openness and appreciation of diversity. This would, however, be far too easy and apolitical. It is clear that hard and functionalist forms of systems thinking dominate within the CGIAR, and that more socio-political perspectives are marginal at best. Most researchers in the CGIAR have a technical and/or bio-physical orientation and training, and while economists are indeed present, the integration of social and political sciences has remained problematic (Thiele *et al.* 2001; Cernea 2005; Dalrymple 2005; Kassam 2006; Kristjanson *et al.* 2009). The core business or comparative advantage of the CGIAR is still considered to be technology development, and for many researchers and research managers, systems research is mainly about integration of different natural science perspectives (e.g. looking at trade-offs between production objectives and natural resource management objectives), or promoting adoption of new technology. While there may be pragmatic reasons to de-politicise research programmes and render agricultural development as a largely technical affair (Li 2011), others argue that the key constraints to achieving development goals are in fact social, institutional and political (Hounkonnou *et al.* 2012; Schut *et al.* 2016b).

As we will argue later, the diverse views on the need of addressing social and institutional constraints, and the lack of a clear and consistent view on systems research, may have weakened the position of systems research within the CGIAR.

Impacts on and in systems: contestation around 'capacity to innovate' as an intermediate development outcome

As part of the reform process, and under pressure of donors, the CGIAR aspires to be an organisation that is accountable and able to demonstrate its development impact. In this context, it has invested considerable resources in the development of a Strategic Results Framework (SRF). Developing this framework involved several rounds of discussion among working groups, CRP leaders, the Independent Science and Partnership Council (ISPC) and the CGIAR Fund Council, as well as several public consultations. An early version of the SRF was published in 2011 (CGIAR 2011a) and contained four strategic System Level Outcomes (SLOs) that were formulated in a largely qualitative manner. In later years, considerable effort went into further detailing and quantifying the SRF. In the process, the number of SLOs was eventually reduced to three, and a range of Intermediate Development Outcomes (IDOs) and sub-IDOs were added (see Figure 5.1 for an outline of the latest version). CRPs are expected to set quantitative targets and make clear how each aspect of their work (alongside a hierarchy of Flagships, Clusters of Activities and Activities) contribute to the realisation of the objectives specified in the framework. Typically, this is done in the form of so-called Impact Pathways or Theories of Change (Alvarez *et al.* 2010) which consist of a narrative and visual diagram that stipulate how research inputs lead to research outcomes, and how

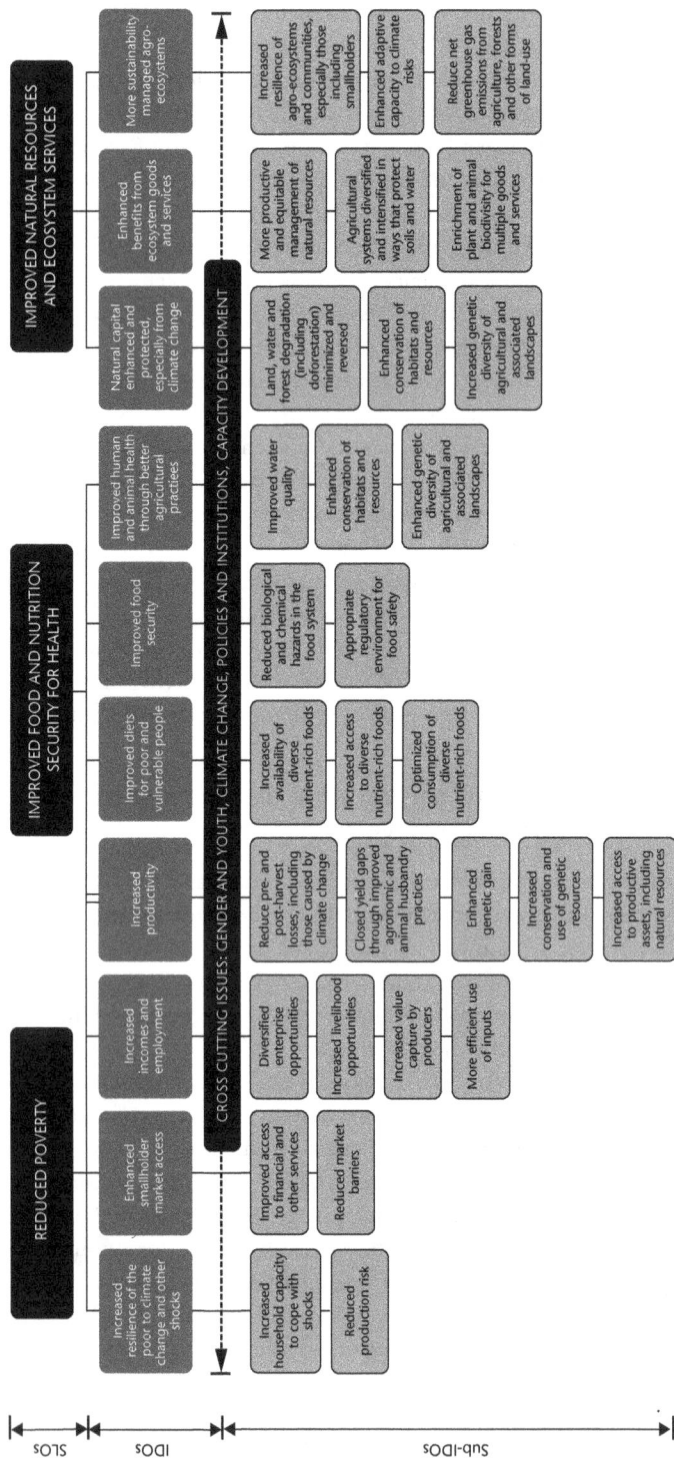

FIGURE 5.1 Strategic Results Framework CGIAR (CGIAR 2015a:15)

these are in turn taken up by 'next-users', leading to development outcomes and eventually impact.

Since the SRF determines the kinds of inputs, outcomes and impacts that are deemed relevant and legitimate, proponents of systems research tried to influence its formulation and evolution during the 2012–2015 period. In particular, the systems CRPs made a case for including an IDO labelled 'Capacity to Innovate'. In several workshops and meetings, researchers from different system CRPs came together to discuss and develop this notion, and overcome different interpretations. Box 5.1 shows how the three systems CRPs eventually described the capacities involved in 'a system's capacity to innovate'.

As Box 5.1 demonstrates, part of the CGIAR systems research community felt that research should be seen as a vehicle for organising collaboration, learning, collective action and coalition formation among interdependent stakeholders. As such, research would generate broader process outcomes rather than simply providing answers to bio-physical or socio-economic research questions. In other words, systems research was not just about doing research 'on systems' and 'for development', but rather it needed to be situated as a strategic lever and catalyst in trajectories of development and change that are inherently institutional and political. To overcome institutional barriers and vested interests, Humidtropics adopted a multi-level approach of working with interconnected local and (sub) national multi-stakeholder platforms. At the local level, 'innovation platforms' would not only work on technical issues, but also identify and address 'institutional constraints' that posed hindrances to the use of promising technology options (see Schut *et al.* 2015a; 2015b for the methodology used). Inspired largely by the Convergence of Sciences programme (Hounkonnou *et al.* 2012), higher-level 'research for development platforms' were to support the development and testing of alternative institutional arrangements and build support coalitions for these in broader policy, development and business networks, thus creating conducive conditions for 'scaling' (Rwampororo *et al.* 2016).

Similarly, it was argued that in order to drive impact, systems research needed to be located in particular place-based settings, and carried out in close collaboration with societal stakeholders and partners. This was framed by the Aquatic Agricultural Systems CRP and others as 'research in development' (Coe *et al.* 2014; Douthwaite *et al.* 2015), and was also one of the reasons why systems programmes paid attentionto embedding research in multi-stakeholder processes, including the establishment of interlinked local level innovation platforms and higher-level platforms (Schut *et al.* 2016a). According to the systems research community, the use of such 'research in development' approaches would simultaneously require considerable change and capacity development in the CGIAR system itself (Leeuwis *et al.* 2014).

As can be noted from Figure 5.1, efforts to introduce 'capacity to innovate' as an IDO were not very successful. Instead, the more conventional notion of 'capacity development' was included in the SRF as a 'cross-cutting issue', along with

BOX 5.1 A SYSTEM'S 'CAPACITY TO INNOVATE' AS DESCRIBED BY THE THREE SYSTEMS CRPS (LEEUWIS *ET AL.* 2014: 5)

Integrated systems are complex wholes in which a range of social and biophysical processes interact across various levels and scales. Reorienting the dynamics of systems in favor of realizing desirable outcomes – for example, intermediate development outcomes – is essentially about changing the way people interact with each other and respond to their changing environment. This requires capabilities at the level of individuals, communities, organizations and networks, and those that have a mandate to catalyze and support innovation processes in society; e.g. international nongovernmental organizations, CGIAR and funding agencies.

Core capacities that are needed at the level of interdependent societal stakeholders:

- the capacity to continuously identify and prioritize problems and opportunities in a dynamic systems environment
- the capacity to take risks, experiment with social and technical options, and assess the trade-offs that arise from these
- the capacity to mobilize resources and form effective support coalitions around promising options and visions for the future
- the capacity to link with others in order to access, share and process relevant information and knowledge in support of the above
- the capacity to collaborate and coordinate with others during the above, and achieve effective concerted action.

In supporting the above, those with a mandate or willingness to catalyze system innovation processes will need to develop the following:

- a conceptual understanding of how change comes about in complex systems and how to intervene effectively
- the ability to orchestrate and facilitate interaction in support of the above
- the ability to inform societal agents and embed research activity in ongoing processes of change.

Together, these capacities form a system's capacity to innovate.

'gender and youth', 'climate change' and 'policies and institutions' (Figure 5.2). While this recognition for 'capacity development' was considered a major achievement by some, much of the meaning implied in the notion of 'a system's capacity to innovate' was lost in the translation as (or integration in) capacity development.

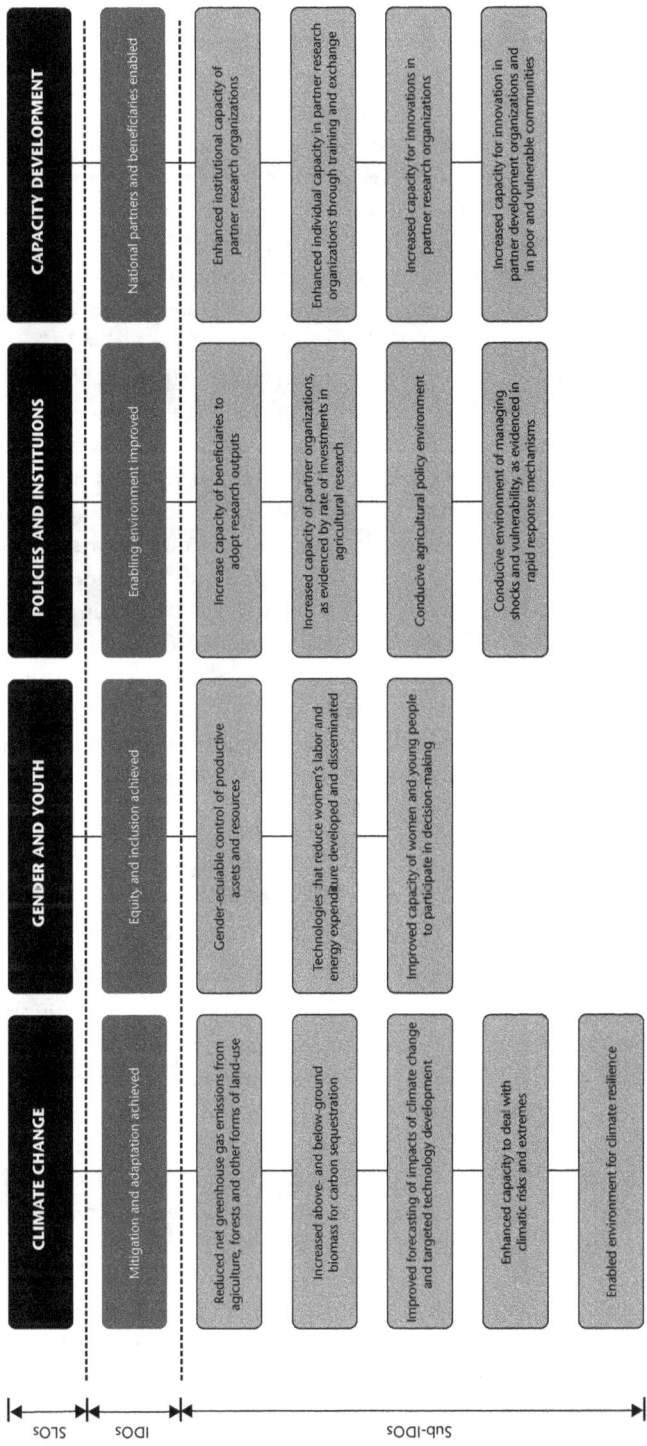

FIGURE 5.2 Cross-cutting themes and outcomes at IDO and sub-IDO level (CGIAR 2015a:23)

From Figure 5.2 it becomes clear that the SRF focusses mainly on capacity development of 'partner research organisations', 'partner development organisations' and 'vulnerable communities'. Neither the CGIAR itself nor actors within the public or private sectors are explicitly mentioned as key parties that may require enhanced capacity to contribute to development. In the explanation of capacity development in the latest SRF (CGIAR 2015a: 22) it is clear that the Consortium emphasises technical capacities like 'data management', 'communication technologies' and 'landscape analysis', and does not mention capacities related to the effective embedding of research in processes of societal change and/or dealing with the socio-political dimensions of development. Such connotations are also largely absent in the description of the SLO 'policies and institutions' (Figure 5.2). Here it is recognised that there is a need to create 'enabling (policy and institutional) environments', but this is to be realised mainly though 'increased capacity of beneficiaries to adopt research outputs' or invest in agricultural research, with research 'providing evidence' as a basis for policy reform (CGIAR 2015a: 22).

What we see in the above is that the SRF embodies a conventional perspective on the role of research in development processes, and that the systems research community within the CGIAR was not successful in institutionalising its alternative perspective. As mentioned earlier, several high level bodies in the CGIAR (including the CGIAR Fund Council, CRP Directors, Consortium Office and ISPC) were involved in formulating the SRF and the process was affected by considerable blurriness and tension about mandates, authority and process. In hindsight, the systems research community probably did not target the right people, and was naive in expecting that a jointly written Programme Brief (Leeuwis *et al.* 2014) would be an influential instrument.

The organisational system of the CRPs: the struggle for control over financial and human resources

Operating along the lines proposed by the systems research community would require long-term investments in multi-stakeholder processes, and the bringing together of interdisciplinary research teams (originating from different Research Centres) around priority challenges and opportunities. Our experience in the CGIAR suggests that the funding and governance arrangements that emerged from the reform process were not very conducive to achieving this. Before the reform, donors channelled funds for the CGIAR through the World Bank, which distributed these to the different Centres. The CRPs were created in part as a mechanism to reduce competition and foster collaboration between different Centres, in line with a broader trend towards donor coordination (Pingali 2010). As part of the reform, donors formed a Fund Council that jointly funds the different CRPs through so-called Window 1 and Window 2 funding (donor funds with different degrees of earmarking to CRPs). In addition, Centres were expected to expand the CRPs through bilateral Windows 3 funding. On a yearly basis, the Fund Council allocated Window 1 and 2 funds to CRPs, and within that to strategic research themes, flagships and clusters of activities. However, over the 2012–2015 period, the Consortium was unable to make funding commitments for longer than one year: the programmes started each year with an indicative budget that only became final in the second half of the year in question, frequently resulting in budget cuts late in the year. This cycle went along with considerable uncertainty and risk among Centres and partners and a need to re-negotiate budgets and activities on a yearly basis. In addition to frustration, this fostered a short-term outlook and culture, quite the opposite of what would be expected in programmes that were formulated initially with a horizon of fifteen years and with long-term transformative ambitions.

Even more significantly, the functioning of at least some programmes was affected by the fact that CRP leaders had in practice limited control over financial and human resources. Since resources were – at the level of intent and rationale – allocated to CRPs, it is not surprising that CRP directors, theme and flagship leaders wished or expected to control these. However, participating CGIAR Centres felt they had a 'share' in the CRP, and observations suggest that they continued to talk about Window 1 and 2 funding as if it was 'core funding'. Moreover, the collaborating Centres held contracts with the lead Centre, while the lead Centre held the CRP contract with the Consortium. Thus, the Directors of Research of various Centres had a great deal of control over financial resources. In practice, therefore, CRP directors and other CRP bodies depended on the willingness and ability of the various Centres to allocate money (and people) to certain activities. Observations from Humidtropics suggest that this institutional arrangement negatively affected the capacity of the CRP to operate as a coherent programme. In particular, it was difficult to mobilise funds and staff time to address issues that emerged from multi-stakeholder platforms in the geographically defined Action Areas since Centres had allocated these resources to other activities. The four Action Area coordinators initially controlled very limited resources from the Window 1 and 2 funds. In 2013 there were no specific operational budget allocations to Action Areas, while in 2014 about US$ 600,000 was allocated to them (amounting to 3.5 per cent of initially projected US$ 17 million budget from Windows 1 and 2). These funds mainly served to cover facilitation and operational costs associated with the platforms.

Initially, it was hardly possible for multi-stakeholder platforms to initiate new research or bring in relevant outside expertise from participating Centres, even if this was warranted by the diagnosis made by platform members (Schut *et al.* 2016a). These problems were recognised by the CRP leadership, which made various efforts to increase budgets that could support research activities initiated by platforms, at the expense of budgets allocated directly to Centres. In an attempt to improve the 'systemness' of the CRP, the director succeeded in increasing the funds under the control of Action Area coordinators to 15 per cent for the year 2015 (US$ 1.88 million of a total projected budget of US$ 12.6 million from Windows 1 and 2). This was despite a significant overall budget cut imposed by the Consortium when compared to 2014. While Centres remained reluctant to further reduce the budget share under their direct control, the sense of crisis created by the budget cuts seemed to foster a climate in which all partners realised that something had to change. The re-allocation of budget was agreed on condition that Centres could participate in demand articulation and project formulation activities in response to issues raised by platforms, and that key Centres might in this way regain some of the lost funds.

Overall, while systemic ambitions in the Humidtropics CRP required effective collaboration among multiple CGIAR and non-CGIAR partners in the Action Areas, our experience was that coordination among partners was difficult as a result of distributed authority and leadership arrangements and prevailing funding mechanisms. Similar observations about counterproductive organisational complexities have been made for the predecessors of the CRPs, suggesting a degree of continuity despite the reform process (Ekboir *et al.* 2013).

The eventual demise of the systems CRPs

After the first cycle of CRPs (2009–2015), the CGIAR decided to work towards a renewed CRP portfolio from 2017 onwards. In May 2015, the Directors General of the Centres met in Windsor (UK) with the Consortium Office and other key actors to reach agreement on a new portfolio and launch a call for new CRP proposals. This meeting took place only a few weeks after the Aquatic Agricultural Systems CRP had received a critical external evaluation (see Birner *et al.* 2015), and amidst considerable pressure from donors to reduce the number of CRPs from 16 to around 8. Reportedly, different options for clustering and combining CRPs were discussed, and eventually only the three systems CRPs, including Humidtropics, were cancelled. At the same time the commodity CRPs re-positioned themselves collectively as 'systems' programmes using terminology like 'maize-based systems', 'rice-based systems', etc. In the eventual call, these commodity programmes were re-labelled 'Agri-food System Programs'. Late in 2015, there was a failed attempt by the Consortium Board to re-open the discussion and work towards eight CRPs, with several variants including the merger of several commodity CRPs (e.g. Rice, Maize, Wheat and Dryland Cereals into 'Cereals') or the re-creation of 'systems' CRPs for different agro-ecological zones (including a CRP for the humid tropics). Eventually, both the Aquatic Agricultural Systems CRP and the Dryland Systems CRP (which also received a critical external evaluation in October 2015, see Merrey *et al.* 2015) were closed down in the beginning of 2016. The Humidtropics CRP was evaluated more positively (CGIAR 2015b; Rwampororo *et al.* 2016) and was allowed to continue until the end of 2016.

In the new CRP portfolio, integration at the geographic level is to be organised through 'site integration plans' with emphasis on a limited number of countries. Different CRPs are expected to align their activities in these countries, and jointly develop and fund these integration plans. At the same time the site integration plans are regarded as the key vehicle for interaction with stakeholders and the realisation of development impact (CGIAR 2015a). However, given the likely absence of a central budget and authoritative leadership, it is difficult to see how the coordination problem identified in the previous section can be avoided.

Analysis and discussion

As can be noted from the above, systems research in the CGIAR remains a highly contested space. There are several dimensions to this.

The nature and place of systems research in the CRP portfolio

First of all, it is important to note that the notion of systems research still has some currency, despite the demise of the three system CRPs. This demonstrates that there is at least some recognition among CGIAR Centres and funders that one cannot focus on single commodities and/or cropping systems only, and that different system boundaries and trade-offs need to be taken into account. We have seen that in the current setting, 'systems' delimitations that emphasise bio-physical, technical and economic dimensions still seem dominant, and that attention to socio-political phenomena and institutional issues is still relatively limited. Similarly, we have seen that – in a context of pressure and increasingly scarce resources – a struggle emerged about where (in which CRPs) systems research should be located and funded. In this struggle, the commodity programmes managed to re-position themselves as the future custodians of systems research at the expense of the original systems CRPs.

Apparently, the original systems programmes did not have sufficient credibility and political clout to carry them through the negotiations and politics that framed the second round of CRPs. Throughout our three years of participant observation in Humidtropics, we often observed that the systems programmes were portrayed as being 'weak'. Criticism was typically formulated along three lines: (a) insufficient clarity about what distinguished them from other programmes; (b) insufficient demonstration of impact and added value; and (c) the place-based approach did not lead to 'international public goods' and duplicated the work of NGOs and national research institutes. Moreover, the evaluations of Aquatic Agricultural Systems and Dryland Systems emphasised limitations in the quality of science produced, underdevelopment of the systems research approaches applied, and poor linkages between conventional research and systems research (Birner *et al.* 2015; Merrey *et al.* 2015). While criticism was certainly warranted one could raise the question whether it was even appropriate to evaluate the systems programmes. Humidtropics, for example, had effectively only run for 2.5 years when the decision to discontinue the programme was taken. One may wonder whether it is realistic to expect clear impact and a coherent approach in such a short time, and immediately following a major re-organisation of the CGIAR system – especially since long-term research involvement has been found to be a key factor in achieving lasting impact (Humphries *et al.* 2015).

Different discourses on CGIAR identity and the role of place-based research in delivering development impact

A key unresolved debate that affects systems research in the CGIAR is around the role and importance of place-based research. This debate is linked to different demands made on the CGIAR, and tensions and contradictions that these pose *vis-a-vis* its mandate and identity. Others have described and analysed this debate in terms of conflicting views on the extent to which the CGIAR should play an active role in securing 'uptake' of technology in situations where national systems have a low capacity in promotion and dissemination (Pinstrup-Anderson 2005; Kamanda 2015). Kamanda (2015: 68) identified two different narratives about how the CGIAR might contribute to impact: a 'Pro-Uptake' narrative and a 'Pro-IPG' narrative. In the former it is argued that worthwhile 'on-the-shelf' technologies do not reach farmers due to failures in delivery (e.g. extension) systems, and that the CGIAR should – in order to achieve impact – play a leading role in facilitating partnerships and capacity development in national innovation systems. In contrast, supporters of the Pro-IPG narrative argue that there is still a lack of suitable technologies, and that the CGIAR should concentrate on the production of international public goods (IPGs) and avoid taking over responsibilities of national governments (Kamanda 2015). While these competing – and both largely linear – discourses were also recognisable within and around the systems CRPs, our experience suggests that there exists another important cleavage in current discussions on systems research which is less about the organisational mandate and responsibility of the CGIAR, and more about the role and nature of research itself. Table 5.2 summarises two co-existing and competing discourses on the role of research in generating impact.

On the one hand, systems researchers working within the 'research in development' discourse tend to argue that societal impacts can only be generated 'in context' and hence that integrative research needs to be carried out in geographically defined areas, and in close collaboration with stakeholders. And since such stakeholders are only likely to engage if they feel the research addresses their needs, proponents argue that stakeholders must to some degree be able to steer the collaborative research effort. Moreover, they tend to believe that the process of doing the research with partners can already generate important benefits and dynamics that are critical to the eventual achievement of impact, regardless of the actual research findings and research outputs. This conviction is essentially based on the idea that there are many social and institutional barriers and disincentives to the uptake of promising technology, and that place-based research processes can play a key role in overcoming these. It is also argued that changing the prevailing incentive systems in society (i.e. fostering institutional change) is a long-term process that cannot be designed, controlled, disciplined or measured through the logic of instruments such as specified impact pathways and the SRF.

On the other hand, the 'research for development' discourse is informed by the view that development impact should be generated by development partners

TABLE 5.2 Key areas of disagreement and contestation constructed in discourses related to place-based systems research in the CGIAR reform process

	Impact through 'research for development' discourse	Impact through 'research in development' discourse
Identity of the CGIAR	A global research organisation that produces international public goods	A global organisation that embeds its research in place-based development trajectories
Major hurdle towards making research contribute to impact in national contexts	Weak 'delivery systems' for disseminating technology	Prevailing institutional set-ups do not provide incentives for technology-uptake
Meta theory on the relation between research and development impact	Research serves to generate and test technical options Research findings are the basis for achieving impact Research phase can be separated from impact phase The pathway from research to impact is largely controllable and predictable Research can generate SLO-type impact within the timespan of a project	Research serves to generate and test technical and institutional options Research process creates conditions for generating impact Research phase cannot be usefully separated from impact phase The pathway from research to impact is largely uncontrollable and unpredictable Research is unlikely to yield SLO-type impact within the timespan of a project
Preferred mode of assessing progress towards impact	Progress towards impact can be usefully assessed through specified impact pathways and the SRF	Specified impact pathways and SRF are not sufficient for assessing progress towards impact Additional indicators are needed to assess changes in societal relations, coalitions, capacities and discourses
Role of place-based development partners and context	Indirect role in steering research through SRF and international consultations Partners integrate insights Centres cannot be trusted in articulating demands with partners (so donors must step in)	Direct role in steering research Partners co-produce insights Centres can be trusted in articulating demands with partners
Modes of systems thinking needed	Emphasis on 'hard' and 'functionalist' systems thinking	'Hard' and 'functionalist' systems thinking is not sufficient. Need to incorporate 'soft', 'critical', 'autopoietic' and/or 'institutional' systems thinking
Preferred organisation of systems research	An add-on to 'commodity CRPs' in the form of site-integration plans and/or a systems Flagship	A separate CRP that integrates insights from multiple Centres and CRPs

(e.g. NGOs and local research institutions), who make use of international public goods including research findings and technology. In this line of thinking, conducting research in response to problems posed by place-based stakeholders is outside the remit of the CG unless it will first and foremost result in the production of IPGs. When asked about the role of members of place-based R4D platforms in steering research, a high ranking officer at the Consortium Office commented in January 2015 that 'the donors regard themselves as those who represent the demand side', and explained further that the donors did not trust that the CGIAR Centres would themselves ensure that research responded to the development agendas of the countries in which they work. In this discourse, the formulation of the SRF, and the organisation of broad stakeholder consultations about this through e.g. the Global Conference on Agricultural Research for Development (GCARD), is seen as a mechanism to represent the demand side (and impose discipline on the wayward tendencies of the researchers). Moreover, placing a strong emphasis on the SRF and the Results-Based Management culture that goes along with it reflects a high level of confidence in the possibility of controlling and predicting the way in which research contributes to development pathways and impacts. The fate of the systems CRPs indicates that the 'research for development' discourse is still dominant among those steering and funding the CGIAR reform process.

Combining the creation of international public goods with engagement in development

We have seen that the notion of international public goods plays a significant role in the struggle around systems research. This notion was extensively discussed within CGIAR circles in the period preceding the 2009–2015 reform (see e.g. CGIAR Science Council 2006; Ryan 2006; Dalrymple 2008; Sagasti and Timmer 2008). Although most of these authors pointed to complexities in defining and operationalising the notion of IPGs in the context of international agricultural research, it has nevertheless become a key anchor for the CGIAR, and a core concept in legitimising its funding. Given that criticism towards the systems CRPs was regularly framed in terms of their inability to produce IPGs, it is interesting to note that during our three years of participant observation in the implementation of systems research we were never party to any discussions about the (problematic) nature of IPGs.

In the scientific literature, Harwood *et al.* (2006: 381) have defined IPGs in the CGIAR context as 'research outputs of knowledge and technology generated through strategic and applied research that are applicable and readily accessible internationally to address generic issues and challenges consistent with CGIAR goals'. In a more popular CGIAR publication IPGs are described as 'products, goods, methods, services, software, knowledge, etc. freely available for use by all' (CGIAR 2011b: 5). In practice, the term is often associated with genetic resources (e.g. new varieties, gene banks), technological designs, methodological approaches, publicly accessible databases and novel insights with a broad relevance. Moreover, the production of IPGs is often judged by the number of publications in international, high impact, and peer-reviewed academic journals. Essentially, the reservations with regard to place-based systems research seem to stem from the assumption that it generates contextually specific insights that are not internationally relevant and applicable. This is an area that merits more exploration and reflection for several reasons.

First, earlier research programmes within the CGIAR and elsewhere showed that it is quite possible to combine action-oriented and development-oriented research with the production of scientific articles in high impact journals (Hounkonnou *et al.* 2012; Struik *et al.* 2014; Devaux *et al.* 2009; Kristjanson *et al.* 2009). This suggests that place-based research can be positioned in relation to scientific knowledge gaps and international debates, and can thus produce IPGs. This would indicate that the tension between development-oriented research and the production of IPGs is not intrinsic.

Second, an obvious comparative advantage of the CGIAR is that it is in a position to do research and compare and synthesise findings across many different contexts. However, without place-based research such comparison and synthesis is impossible and this comparative advantage cannot be realised.

Third, it is clear that, eventually, all research takes place in a specific context (even if this is an experimental farm, a laboratory or the set of assumptions that frame an econometric model), and that the features of that context determine to a considerable degree the validity of the findings in other contexts. Hence, it is a mistake to think that research can take place outside a specific context. Rather, to enhance the relevance and public value of development-oriented research, it is essential to make deliberate and well-motivated choices about the contexts that are chosen.

Fourth, based on the conventional definition of international public goods one may also argue that an IPG has not been produced unless it has indeed been made 'applicable and readily accessible internationally'. Experience tells us that making insights, technologies, institutional designs or services that were relevant in one context also applicable and accessible in another, requires special efforts and engagement that go beyond placing them on a central website or publishing them in a journal (see Glover *et al.* this volume). Thus, one could argue that an ambition to produce IPGs inherently requires complementary place-based research activities. In more theoretical terms: we may consider that 'a good' does not become 'internationally public' through its static features, but rather through an active and engaged process of making it accessible and applicable.

Fifth, we may want to broaden the definition of what constitutes an IPG. Currently, the emphasis is on the production of goods and services that are usable for many nations at the same time, but one could also define them as all research-based goods and services that cannot be delivered by national parties in a specific context. That is: as all research-based services that add value to national systems. Recent literature on the role of research and researchers in society has emphasised that researchers may usefully play broader roles than knowledge production or technology design in efforts to achieve societal impact. Such roles include various forms of innovation management and innovation intermediation (Klerkx *et al.* 2009; Schut *et al.* 2014; McNie *et al.* 2016) that tend to be poorly developed in many countries. An 'added value' based definition of IPGs would allow CGIAR researchers to legitimately play such process roles in contexts where national innovation capacity needs to be strengthened.

In all, we see that the contradictions that are constructed in the discussion on systems research are to some degree artificial and that they may well be bridged.

As indicated in the introduction, key objectives of the CGIAR reform were to make the system more demand driven, development relevant and accountable. Logical questions to ask then are: Whose demands? And accountable to whom? From the experience with systems research, it has become clear that the demands that donors formulate on the basis of international consultations play a central role in the CGIAR, and that accountability is mainly framed and designed in terms of upward accountability to donors, the Consortium Office and the ISPC, with the SRF as an important point of reference. In line with this architecture, Leeuwis experienced that the specific demands and priorities of national governments, development partners and private sector stakeholders were never the explicit starting point for discussion and decision-making during higher-level management meetings. In Action Areas such demands and priorities did play a role, but given the prevailing modes of allocating funds and human resources, it was not always easy to accommodate these (see Schut *et al.* 2016a; 2016b). Arguably, this weak orientation towards those who are eventually responsible for effectuating change is unlikely to foster the type of interaction that is found to be conducive to making research contribute to innovation in society (Klerkx and Leeuwis 2008; Roux *et al.* 2010).

Conclusions

Within the CGIAR, socio-political and institutional dimensions of agricultural and rural change receive less attention than technical dimensions, reflecting the CGIAR's particular history, identity and politics. In part, these politics relate to a wish to obtain and control funding in an environment of budget cuts and a donor community that increasingly demands that research organisations promise and provide evidence of development impact over relatively short time horizons. There is no evidence from within the CGIAR or beyond, that making and assessing such promises is either realistic or productive. Moreover, while others have pointed to different views regarding the desirability of the CGIAR engaging in place-based partnerships to stimulate uptake of technology, our experience suggests that there exists a deeper cleavage in the form of competing meta-level theories of change on how different types of research may help foster development. While some see place-based systems research with stakeholders as a promising way to catalyse change, others see it as a distraction from the CGIAR's core business of producing international public goods. These divergent views are played out in the struggle for resources, with the latter perspective holding sway. While the construction of an unbridgeable gulf between place-based systems research and the delivery of international public goods has been effective in de-legitimising the systems CRPs, we have argued that there need not be a contradiction between the two, and there may be potential for synergy.

In view of an agenda for further research and action, we suggest moving beyond discussions on whether systems research should be maintained, and where it should be located institutionally. Asserting that systems research remains an important part of the new Agri-food Systems CRPs is one thing, but grappling with questions about the types of systems research that are needed, and how they may be operationalised to enhance development impact is quite another. In relation to this, Leeuwis and Wigboldus (2017) suggest making systems research more action oriented through the systemic evaluation of combined technical and socio-institutional experiments. In addition, it is important that the CGIAR reflects critically on its own capacity to work in a systemic way. In essence, the experience with the three systems CRPs demonstrates that the CGIAR environment was not conducive to implementing systems research. This was because (a) there was no general agreement about how different modes of systems thinking should be integrated, (b) expertise from innovation studies and social sciences remained scarce, (c) researcher roles and impact assessment approaches relevant to systems research were not congruent with the dominant culture (including the SRF), and (d) the way of funding and governing research posed significant hurdles to systems research. As these issues are not easily resolved, the new CRPs and site-integration efforts may continue to yield disappointing results (see also Schut *et al.* 2016a, 2016b).

With regard to the issue of impact on or in systems, there is a need to better understand the (in)compatibilities between systems research and the way in which donors think about and measure impact. Novel frameworks and approaches that can provide credible alternatives are badly needed (see Leeuwis 2013; Arkesteijn *et al.* 2015; Glover *et al.* 2016). Lastly, in relation to the organisation of CRPs, there is a need to further unravel accountability relationships and politics in research agenda setting and funding mechanisms. Applying principle-agent theory to these issues may not only shed more clarity on how complex webs of relationships enhance or constrain impact of research on and in systems, but also offer practical inspiration for delegating authority in research funding to networks of stakeholders (Braun 2003; Klerkx and Leeuwis 2008).

Acknowledgement

We would like to acknowledge Humidtropics and the CGIAR Fund Donors (www.cgiar.org/who-we-are/cgiar-fund/fund-donors-2/) for their provision of funding without which this research could not have been possible.

Note

1   See: http://humidtropics.cgiar.org/wp-content/uploads/downloads/2015/03/Detailed-Conference-Program.pdf

6

ONE STEP FORWARD, TWO STEPS BACK IN FARMER KNOWLEDGE EXCHANGE

'Scaling up' as Fordist replication in drag

William G. Moseley

Introduction

A few summers ago I spent several hours in an airline lounge in Paris as I had a long layover between flights from the United States to Mali where I was to do preliminary fieldwork for a research project on the New Green Revolution approach in Africa. Although I had worked on and off in Mali since 1987, I was very keen to return because it was my first time back since a coup d'état in 2012. In addition to work, I wanted to see old friends and I was very curious to see what had changed.

As I settled into a bank of workspaces to call a number of colleagues and friends in Mali before I arrived, a thirty-something woman sat down next to me and similarly set about her work of making calls and typing on her laptop. Try as I might to ignore her voice, I could not help but overhear snippets of the dozen or so phone calls she would make over the next two to three hours. It became clear, by her accent, that she was American, likely well educated and that she was director of an NGO working in southern Mali on agricultural issues. It was June (still early in the rainy season) and she was communicating with various staff about getting agricultural inputs out to farmers in a timely fashion. She talked to one staff member about soliciting funds from a private sector donor, Orange, the French cell phone company. This was for support to bring their project to 'scale'. She actually spoke repeatedly about scale in her various calls with staff. It seemed as if it was almost an obsession for, in her view, all would be for naught if they could not bring this project to scale. The other aspect that became obvious in these conversations was that she travelled a lot. She had just been in the US for a conference at Harvard. She would then be in Mali for a week before heading to a meeting in East Africa and then up to Europe.

So what have we learned here (other than not to speak too loudly on the phone while in airports)? My new friend at the airport, whom I would never actually meet but subsequently would learn a lot more about, was part of a new breed of development workers. She had a business degree from an American Ivy league school, was passionate about international issues (particularly hunger), carefully framed the development organization she founded as a social enterprise (not an NGO), and believed that her approach was new, different and would succeed where others had failed. She also represents a particular imaginary, a global jet setting elite who directs projects from international airport lounges. Her approach to agriculture, frequently called the New Green Revolution for Africa, involving improved seeds, pesticides, fertilisers, the commercialization of agriculture, and business-savvy development professionals, is also the subject of my study.

To be more specific, I am exploring the 'scaling up' approach to extension embedded in the New Green Revolution for Africa's emphasis on value chain construction. I seek to understand how this approach is similar to, or different from, previous methods, and how this tactic resolves or accentuates agro-development challenges related to gender and participation. This approach to knowledge dissemination is explored using case studies in Mali. My findings are based on 30 years of work and research experience in Mali in which I have informally observed NGO and government approaches to extension. In particular, I rely on a set of semi-structured interviews from June 2014 with government and NGO representatives about the New Green Revolution for Africa approach in southern Mali. This included observation of extension staff working for a social enterprise (inspired by this paradigm) in southern Mali.

Context in the literature

This chapter is informed by ideas and theories from cultural and political ecology, as well as contested agronomy. Certain traditions in cultural ecology have sought to validate traditional agricultural practices in the tropics as rational (Moseley *et al.* 2013). Cultural ecology was, in part, a reaction to colonial modes of management which were grounded in Eurocentrism. As Richards notes (1985: 11): 'French, Belgian and British colonialists, convinced of their own intellectual and cultural superiority, failed to understand both how particular and place-bound were their own principles of environmental resource management, and the extent to which many of the characteristic practices of African farmers and pastoralists were effective responses to the highly specific challenges posed by the African environment.' Scott (2009) has made similar points about historically and regionally situated western modes of understanding being pitched as universals. Political ecology built on the insights of cultural ecology about local agricultural practices, but inserted the need to understand these strategies as nested in political economy dynamics at the local, national, regional and global scales (Moseley *et al.* 2013). As such, one cannot understand changes in agriculture, and orientation of agricultural extension, without understanding political economy. Post structural political ecologists have also written extensively on environmental narratives, or the discursive framing of environmental management issues as influenced by power dynamics (Leach and Mearns 1996; Robbins 2012).

This chapter is also informed by the discussion of contested agronomy. As Sumberg, Thompson and Woodhouse (2012) and Ross (2014) have suggested, development-oriented agronomy, originally tropical agronomy, has its roots in Europe and was driven by the need for colonial powers to capture and modify tropical crops, soils and farming practices as fuel for European economic expansion. In the post-colonial period, Sumberg, Thompson and Woodhouse (2012, 2013) further argue that development-oriented agronomy has been influenced by the neoliberal project, the environmental agenda and the participation agenda. When thinking specifically about agricultural extension practices and development-oriented agronomy, one could argue that there is both complementarity and competition between the participation agenda and neoliberal project. In the first instance, it could be that the participation agenda is consistent with the neoliberal project because increased decentralization and local-level decision making may render national governments less important (Mann 2015). In the second situation, however, increased participation may lead to greater expression of local-level particularities, both social and ecological (Richards 1985). Such local particularities may rub up against the need to standardize in capitalist oriented agriculture (McMichael 1997).

Understanding knowledge politics is also central to contested agronomy (Andersson and Sumberg 2017). As Vanloqueren and Baret (2009) note, an analysis of knowledge politics in agronomy helps explain why some technologies or development pathways are privileged over others. I would further argue that embedded in knowledge politics are certain ways of knowing and doing. With respect to the New Green Revolution for Africa, the influence of a business-oriented donor community (Schurman and Munro 2014) has likely resulted in certain modes of practice in agricultural extension.

The evolution of extension in tropical agriculture and development

Jones and Garforth (1997) have defined agricultural extension as including 'activities which seek to enlarge and improve the abilities of farm people to adopt more appropriate and often new practices and to adjust to changing conditions and societal needs'. While various forms of agricultural extension have been around for centuries, more modern versions of this approach likely date to the mid-nineteenth century in the UK. It was at this time that both Oxford and Cambridge universities began to think about serving populations in surrounding communities. While agriculture was not the initial focus of these efforts, it began to receive more attention by the 1890s.

Developments in the UK influenced thinking in the United States and spread to newly created land grant universities with explicit public missions. This happened to coincide with a political movement in the United States known as agrarian populism that was organized by small and medium farmers against urban-based financial speculators (Schneiberg *et al.* 2008). According to Richards (1985: 16), in this era 'American extension agents were seen, initially, as employees of the farming community, not as agents of a centralized scientific bureaucracy. A priority for a number of early extension services in the United States was to communicate farmers' needs to researchers, not to disseminate scientific findings to potential users. Under a populist rubric extension workers were truly "agents," rather than the educators, communicators and even salesmen, they have since become.'

This is not the extension model that comes to Africa in the colonial era, through the activities of some colonial agricultural research stations as well as missionary outposts in the early twentieth century (Jones and Garforth 1997). Here there was little interest in working with African understandings of tropical agriculture, but rather the focus was on getting Africans to adopt European practices and/or to grow crops for the commercial market. In many cases, local, subsistence-based systems were seen as a problem and frequently framed as under productive and environmentally destructive (Beinart 1984). In West Africa, a combination of taxation, coercion and extension was used to develop cash crop farming in the region in the colonial period, a process often met with considerable local resistance and agency (Bassett 2001).

In Mali, the French colonial authorities were mainly interested in working with local farmers to produce cotton. They initially focused their efforts on irrigated cotton in the Office du Niger (middle Niger valley around Segou) but met with limited success. Cotton production for the external market expanded in the 1950s when the French began to focus on the promotion of rain-fed cotton (as opposed to irrigated varieties) in the southern third of Mali, mainly the Sikasso, southern Segou, and southern Koolikoro regions. The French parastatal Compagnie Française pour le Développement des Fibres Textiles (CFDT), or the French Company for the Development of Textile Fibres, was responsible for facilitating cotton production in southern Mali (Roberts 1996; Bingen 1998). Early French work with farmers in Mali was highly coercive (Becker 1994), eventually moving to a model of price incentives to encourage production (Roberts 1996). In both cases, there was a consistent effort to modernize agriculture along the lines of European agricultural methods and systems (van Beusekom 1997).

In the post-colonial period, the 'Training and Visit' extension method, or T&V, prevailed throughout most of the 1970s, '80s and '90s in many African countries (Anderson 2006). T&V is an extension methodology wherein the agent meets repeatedly with select groups of contact farmers who in turn are to disseminate what they learn to others in the community. It is in part based on models of diffusion, i.e. how ideas and practices spread across a landscape in space and time. This was the dominant approach to agricultural extension when I worked and did research in Mali between 1987 and 1995. In interviews, agricultural extension agents, working for the Office de la Haute Vallée du Niger (OHVN), often characterized their relationship with farmers as that of a teacher and his students, making it clear who had the knowledge to share (Moseley 1993). Ideally agents would discuss problems with farmers, and then return over the coming months to share a number of technical packages with the farmers as a way to address these problems. One of the challenges I noted at the time was the power differential between extension agents and local farmers, and the fact that book knowledge was valued over local knowledge (Moseley 1996). In southern Mali, this approach to extension was overshadowed by an over-arching preoccupation with growing more cotton (Moseley 2008). As such, extension efforts were almost exclusively focused on male farmers (the target audience for cotton growing) who were taught techniques developed on research stations (see Figure 6.1). While farmers often deviated from these prescriptions when extension agents were not present, based on their own knowledge and particular constraints, the actual T&V sessions were run in a top-down, proscriptive manner. As others have discussed, the Malian state and its agricultural extension efforts, weakened by neoliberal economic reform, had become the hostage of a single commodity (cotton) whose sale paid most of the bills (Keeley and Scoones 2003).

With an increasing awareness of local knowledge, prompted in part by cultural and political ecology in anthropology and geography, more participatory research and extension methodologies began to appear from the late 1980s to the 1990s. The most popular of these was a suite of techniques developed and made popular in several works by Robert Chambers (e.g. 2008), initially known as Rapid Rural

FIGURE 6.1 Field day in southern Mali in the late 1980s. A group of (all male) farmers being taught techniques for growing cotton by OHVN extension staff. Photo by author.

Appraisal and then Participatory Rural Appraisal. This approach, using participatory techniques such as community mapping, proportional piling, transect walks, and Venn diagrams (see Figure 6.2), sought to make the research process more transparent, participatory and accessible to illiterate subjects, and – critically – open to local insights and input. Central to this method was a core belief that local knowledge could make a critical contribution to development initiatives and therefore research methodologies were needed to make this know-how legible to outsiders (Chambers 1994).

While this extension and research approach never made significant inroads into government extension work in Mali, it did have a big impact on NGOs and their practice during this period. While working for Save the Children (UK) in central Mali in the mid-1990s, I saw first-hand how participatory methodologies were incorporated into extension work and assessments (Moseley *et al.* 1994; Moseley 1995, 2007). One of the more remarkable initiatives using these approaches at the time was the NGO World Neighbors in the Segou region of central Mali (Gubbels 1994). While this group sought to improve food production, food security and

FIGURE 6.2 Participatory Rural Appraisal, using progressive piling, with female farmers in Zimbabwe in mid-1990s. Photo by author.

nutrition in rural communities (especially among women), it started with the assumption that local insights and know-how were critical to this process. According to Gubbels (1994), the idea was to train 'peasant farmer experimenters' in an approach and method for comparing new technologies with existing practices. The steps in this process were described as follows: 1) farmer diagnosis of agricultural problems; 2) helping communities identify potential innovation; 3) community selection of innovations to test; 4) testing of new technologies; 5) community evaluation of results; and 6) community-managed extension of successful innovations. World Neighbors claimed that this approach was enormously successful, leading to series of innovations that blended local and outside know-how and technologies – and an outcome similar to that advocated by Richards (1985). The programme also claimed success in engaging with and involving women, a process which, it was claimed, eventually led to significant household food security and nutrition gains (Gubbels 1994). The programme continues to function today with limited support from Groundswell International,[1] a successor to World Neighbors.

The New Green Revolution for Africa and reversals in agricultural extension

The New Green Revolution for Africa starts with the basic premise that the first Green Revolution of the 1960s and 1970s largely bypassed Africa, and that Africa now needs its own Green Revolution to jump start development and feed its growing population.[2] As with the previous Green Revolution, the new one involves the use of improved seeds, fertilisers and pesticides to boost crop production (Annan 2007; Toenniessen et al. 2008). However, the New Green Revolution differs from the previous one in a number of ways. Unlike the first iteration (which was largely limited to cash crops and male farmers in the African context), this new one is more focused on food crops and the need to reach female farmers, who are the majority of agriculturalists in many areas of the African continent. Proponents further argue that this new approach is best adopted within the context of global value chains – that is, African farmers need to be integrated into chains of input suppliers, processors and transformers of agricultural products, and retailers (Moseley et al. 2015).

According to Hellin and Meijer (2006: 4), 'a value chain can be defined as the full range of activities which are required to bring a product or service from conception, through the different phases of production, delivery to final customers, and final disposal after use'. The Alliance for a Green Revolution in Africa (AGRA) conceptualizes value chains as the sequence of activities focused on the transformation of a crop from the farm to national levels. At the farm level, the value chain framework promotes the 'upgrading' of agricultural production and food security by integrating farmers into new input and output markets in order to increase production, sales, and income. At the national level, the value chain approach seeks to transform the structure of national economies by promoting investments in agro-processing to add value to products previously exported as unfinished goods. Figure 6.3 depicts these different steps for the AGRA project from inputs, to farming, processing, distribution, and trading.

This approach also involves a new and unprecedented level of public–private partnerships, as donors work to increase the penetration of the private sector and build links between African farmers, input suppliers, agro-dealers, agro-processors and retailers (Hartmann 2012). The private sector, in turn, lauds this new approach and supports it through rhetoric, active involvement in sales, as well as philanthropic activity (e.g. Page 2012). Along with private sector involvement, has come greater emphasis on the need to 'scale-up' after an initial pilot phase – small is no longer beautiful (Schumacher 1973). This last shift has significant implications for the practice of agricultural extension.

Scaling in this context has two meanings. First and foremost, it denotes mass replication akin to assembly-line or Fordist reproduction in older industrial terms. It also suggests the need to jump from local, to national, regional and international scales by integrating small producers into value chains that cross these levels (Hartmann 2012). As I will discuss below, local knowledge returns to the back seat in such a process (as compared to its relative import during the participatory development era, at least among NGOs) and local control is also lost as farmers are more closely integrated into global supply chains. While African farmers have long been integrated into global supply chains for some cash crops, such as cotton (Moseley and Gray 2008) and cacao (Ryan 2011), this is a relatively new phenomenon for food crops.

While not always stated explicitly, local technology and know-how are almost always framed as inadequate in the New Green Revolution for Africa literature, and those who claim otherwise are frequently dismissed as naive. Norman Borlaug, the father of the first Green Revolution, wrote in one of his final publications (Borlaug and Dowswell 1995: 123): 'Some ... contend that small-scale peasant food producers can be lifted out of poverty without the use of modern agricultural inputs ... They envisage soil fertility strategies based on organic fertilisers, farmer-bred and maintained indigenous varieties, biological or mechanical control of weeds, diseases and pests ... In our experience, small-scale farmers are loath to

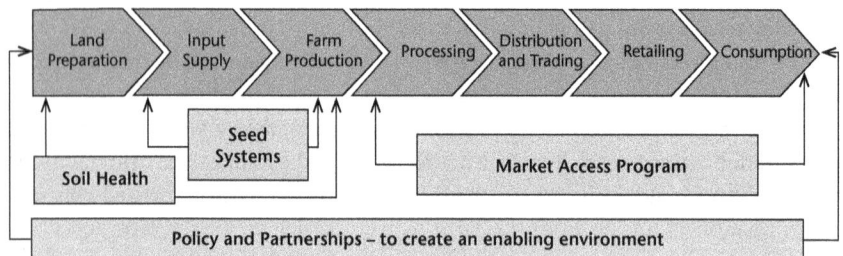

FIGURE 6.3 Alliance for a Green Revolution in Africa (AGRA)'s value chain approach for agricultural products. *Source*: Adapted from Toenniessen *et al.* (2008). Used with permission.

adopt such "low-input, low-output" technologies.' Furthermore, agroecology is not imagined or accepted as a basis for improved production (also see Sumberg *et al.* 2013).

Given the assumption, or bias, that farmers need exogenous technology and know-how to become more productive, extension has returned to older models seen during the colonial and T&V eras that emphasize a more top-down approach to knowledge dissemination. Furthermore, given the increasing presence of social enterprises as vehicles for development, as well as public–private collaborations, the business approach to development has conflicted in other ways with participatory models of extension emphasizing local knowledge. First and foremost, there is no money to be made in leveraging local knowledge. If a social enterprise needs a revenue stream, then something needs to be sold to the farmer-cum-customer. Secondly, in order to be financially viable, social enterprises need to increasingly reach large numbers of customer-farmers in order to be efficient and drive down costs. Lastly, the increasing involvement of business-oriented philanthropists in African agricultural development has resulted in a growing emphasis on metrics to demonstrate impacts and growth.

In June 2014, I spent several days visiting a social enterprise in Mali actively involved with the New Green Revolution for Africa (whose director I briefly discussed in the introduction to this chapter). This organization, which I will henceforth refer to using the pseudonym 'YourFarmer', started working in Mali in 2012. They operate in an area south of the capital city Bamako and introduce technical packages for improved maize and groundnut production. They are also part of a growing number of organizations working in southern Mali to improve food production. For example, Mali is a 'Portfolio 1' country for AGRA, a 'focus country' for USAID's Feed the Future programme, and the recipient of the World Bank's 'Fostering Agricultural Productivity Project'. What these groups hold in common is a core belief that with the right technologies and know-how, southern Mali could become a bread basket for the entire region. YourFarmer believes that farmers in southern Mali are ripe for a transition to high external input agriculture. According to the group's website: 'Malian farmers, if connected to the right inputs, markets and affordable financial services, can get out of poverty, and help feed a growing country and region in the process.' The group's website also discusses the importance of scaling, a process needed to ensure the growing efficiency of their operations. 'As [YourFarmer] scales, we've generated some important lessons that influence how we operate. Lesson 1: Each person we add should be more efficient and lower our costs over time. We are aiming to serve over 450 farmers per field agent by 2022, up from 118 farmers in 2013.'

YourFarmer has developed an innovative savings plan involving the use of cell phones. Over the dry season, male and female farmers save money, purchase cards for a range of values from a network of vendors, and then using a client ID code, are able to upload their savings via a cell phone to a remote account kept by YourFarmer. This savings plan avoids the credit problems associated with some other high external input agricultural programmes. Farmers may purchase packages (improved seeds and inputs) for 1/16th to 1 ha of groundnuts (mostly women) and 1/8th to 5 ha of maize (mostly men). YourFarmer then comes to the village at the beginning of the agricultural season with the seeds, fertilisers and herbicides for each farmer. Part of the cost also includes training from YourFarmer staff.

I was able to observe training by YourFarmer staff in a few of the communities where they work – which also happens to be an area where I have undertaken long-term research. Much of the training involves transmitting a highly prescribed approach to planting and using associated inputs. One of the improved practices that YourFarmer is introducing is the micro-dosing of fertiliser. The basic idea is that one should use a small amount of fertiliser under each planted seed to maximize efficiency (Aune and Bationo 2008). In practical terms, this means digging a hole, depositing fertiliser, adding a little soil, depositing the seed and then covering the seed with soil before moving on to repeat the process (see Figure 6.4). This contrasts with a traditional method of going down a row, planting seeds and then dropping fertiliser in spaces equidistant between each seed. In informal conversations with participating farmers (whom I knew from previous work in the area), several complained that the micro-dosing approach was, in their view, overly labour intensive. They understood the benefits, but just didn't have the time, given other farming demands. Interestingly, YourFarmer staff also complained that farmers just didn't understand the benefits of micro-dosing. At YourFarmer headquarters in Bamako multiple meetings were held in which micro-dosing was framed as an

FIGURE 6.4 Farmers micro-dosing fertiliser in Southern Mali. Photo by author.

educational problem. The response was to change the training to better articulate the benefits of micro-dosing. Labour constraints were not discussed because farmer input was not central to YourFarmer's model of agricultural extension.

The other new extension agents on the ground in Mali are stationary and mobile agrodealers who sell inputs to farmers. For example, the local fertiliser company Toguna, which is an affiliate that mixes for the international corporation Monsanto, has a network of sales agents that work directly with farmers to get them inorganic or mineral fertilisers (often NPK or urea). The entrance of such private sector actors has been greatly encouraged by the donor community and proponents of the New Green Revolution for Africa. While Toguna agents do far less extension than YourFarmer, they do dispense a large amount of technical advice narrowly focused on the use of their products.

Conclusion

Agricultural extension has shifted and evolved over time in Africa. Little has been written about agricultural extension under the New Green Revolution for Africa, with its emphasis on value chains, public–private partnerships and scaling. Using southern Mali as a case study site, this chapter explored agricultural extension under this new paradigm in relation to past approaches during the French colonial, T&V, and participatory eras. While agricultural extension undertaken by NGOs arguably became more of a two-way dialogue between farmers and researchers in the 1990s participatory era, this has all but disappeared in the contemporary period dominated by the New Green Revolution for Africa. Given the dominant assumption that African farmers need new technology and know-how to become more productive, extension has returned to older models seen during the colonial and T&V eras that emphasize a top-down approach to knowledge dissemination. Furthermore, given the increasing presence of social enterprises as vehicles for development, the business approach to development also conflicts in other ways with participatory models of extension emphasizing local knowledge. First, there is no money to be made in leveraging local knowledge. Because social enterprises need revenue streams, goods or services must be sold to farmers. Second, in order to be financially viable, social enterprises need to increasingly reach large numbers of customer-farmers in order to be efficient and drive down costs. Lastly, the increasing involvement of business-oriented philanthropists in African agricultural development has resulted in a growing emphasis on metrics to demonstrate impacts and growth. This push for efficiency and numbers means that the particularities of place, participation, sensitivity to gender and power differentials, local knowledge or two-way dialogue are all but absent. As such, the impact of scaling on agricultural extension means that the New Green Revolution is only masquerading as gender sensitive and participatory. The above also suggests that knowledge politics, and the primacy of Western agronomic knowledge, is central to recent changes in agricultural extension.

Notes

1 Groundswell International works with communities and organizations in Africa, Asia, and Latin America to spread agroecological farming practices, farmer innovation, farmer-to-farmer extension and community health (http://www.groundswell international.org/our-story/).

2 The truth is that the first Green Revolution did not entirely bypass Africa. It did appear in a more limited way and had a significant influence on the production of certain cash crops such as cotton, cacao and coffee. It also influenced the production of some food crops in certain areas, such as maize in Zimbabwe and rice in The Gambia (Carney 2008).

7

WHEN THE SOLUTION BECAME A PROBLEM

Strategies in the reform of agricultural extension in Uganda

Patience B. Rwamigisa, Paul Kibwika, Frank B. Matsiko, Margaret N. Mangheni and Regina Birner

Introduction

Many developing countries are grappling with the challenge of reforming their agricultural extension systems so that they are better able to contribute to the transformation of the agricultural sector. Such reforms have targeted organisational and institutional arrangements, methodologies, financing arrangements and delivery approaches (Akinnagbe and Ajayi 2010). The Neuchâtel Initiative, a donor platform for agricultural extension, proposed a reform paradigm, the core tenets of which included private sector participation, farmer empowerment, demand-driven approaches and decentralisation (Neuchâtel Group 1999, 2002). In line with these tenets, Uganda embarked on a radical reform of its national agricultural extension system which culminated in the establishment of the National Agricultural Advisory Services (NAADS). This reform was instituted through an Act of Parliament (NAADS Act 2001).

The NAADS reform was meant to move away from what was portrayed as a traditional top-down, government-led extension service, based on the Training & Visit model, and toward a privatised, demand-led delivery system (MAAIF 2000; Semana 2002). Under NAADS, farmers were expected to articulate their own knowledge and technology requirements; but even more radically, to own and control the extension service (NAADS Act 2001; Parkinson 2009; Kjaer and Joughin 2012). This vision fully complied with all the principles of demand-driven extension (Chapman and Tripp 2003). However, despite a high level of investment and radical institutional reforms, the NAADS programme ultimately failed to realise its intended objectives (Benin *et al.* 2012).

Different explanations have been advanced to explain this failure, including: political and elite capture (Kjaer and Joughin 2012.); lack of commitment by government (World Bank 2010b); ambivalence (Parkinson 2009) and/or limited understanding on the part of farmers (Musemakweri 2007); limited capacity of private service providers (Mangheni *et al.* 2003; Obaa *et al.* 2005); and an overly ambitious expansion of the programme (World Bank 2010b).

The failure to create a sense of ownership among key policy and institutional actors during the design and implementation of the reform was fundamental in the failure of the NAADS programme (Kjaer and Joughin 2012). These actors comprised two groups. The first subscribed to a 'radical' reform approach and advocated for the total overhaul of the existing structures. The intention was to change the mind-set of the extension workers and managers towards a more efficient, farmer-oriented and performance-based system. The overriding principle was to create a system that incentivised efficiency and effectiveness by bringing in the private sector. Those who subscribed to this view were mainly staff members of the Ministry of Finance, Planning and Economic Development and development partners led by the World Bank. They believed that such a reform was not possible within the existing institutional setup of the Ministry of Agriculture, Animal Industry and Fisheries (MAAIF). We refer to this grouping as the 'radical reform coalition'.

The second group of actors subscribed to a gradual or incremental approach to reform, with a focus on changes and adaptations to existing systems to make them more efficient and accountable. This view was held largely by actors from MAAIF, local government, the National Agricultural Research Organization (NARO), academic institutions and farmer organisations. Together we refer to them as the 'gradual reform coalition'.

The presence of these two opposing coalitions meant that the reform process was designed and implemented in a highly competitive environment that fostered a 'winner takes all mentality' (Rwamigisa *et al.* 2012). This chapter explores the strategies used by the two coalitions and asks how these strategies affected key decisions and outcomes. The chapter aims to contribute to the emerging literature that attempts to explain the failure of agricultural extension reform in Africa. Much contestation centred on the approach the radical reform coalition employed to overcome bureaucratic resistance. Yet domestically, it was perceived as an imposition by donors and outside experts, and exclusive in nature. The tension between donors and (at least some) domestic policy actors persisted throughout the reform process. Despite the apparent strength of the radical reform coalition, the reform ultimately failed because resistance to the reform process was never effectively neutralised.

Conceptual framework and methodology

This analysis makes use of the Advocacy Coalition Framework (ACF) developed by Sabatier and Jenkins-Smith (1993). The ACF recognises that for policy issues that are contested, individuals and groups often come together in 'advocacy coalitions' that coalesce around deeply held beliefs about how the world works. Such advocacy coalitions 'substantively engage in a nontrivial degree of coordinated activities over time' to pursue their preferred policy options (Sabatier and Jenkins-Smith 1993: 18).

In pursuing policy options, coalitions mobilise different types of resources to gain political advantage. Resources can consist of financial resources, human resources, and social networks. The degree of a coalition's influence depends largely on its capacity to mobilise these resources in order to create 'political capital' through activities including stakeholder mobilisation, lobbying, securing support from international actors and use of scientific evidence (Figure 7.1). In this framework, the outcome of the policy process depends on the capacity of different coalitions to build and use political capital.

Data collection started in 2007, approximately 10 years after the initiation of the reform process. A review of documents and initial interviews with officials in MAAIF served to map the policy landscape and provided an overview of the process that led to the creation of NAADS. Based on this initial mapping and a snow-ball sampling approach, 56 semi-structured interviews were conducted with respondents drawn from 14 institutions constituting both state and non-state actors. Participant observation, informal interactions with key policy actors and reviews of internal and published documentation provided additional sources of data.

For each interview, detailed notes were made, including verbatim quotes on key issues. In analysing the interview notes, methods of discourse analysis as

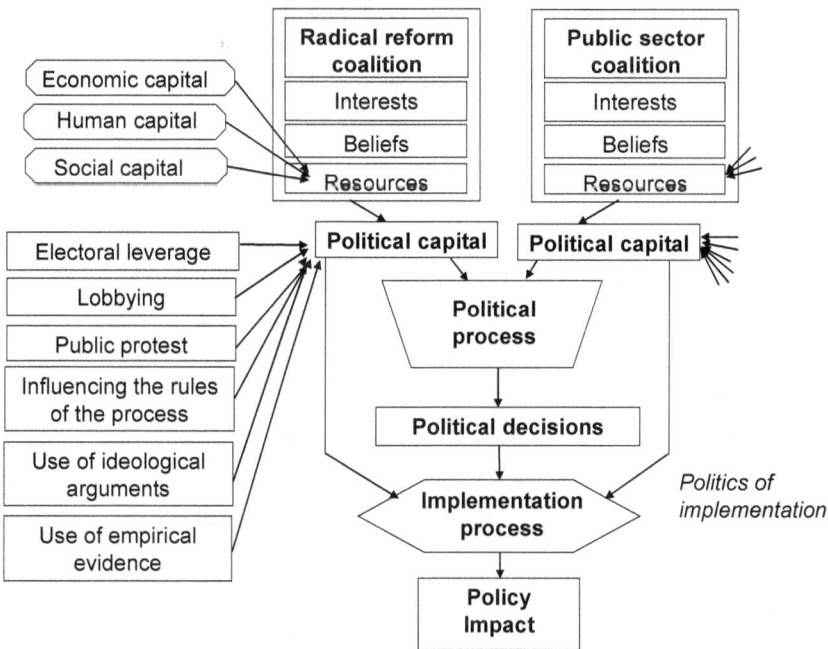

Source: Birner and Resnick (2010, p. 1448), adapted from Sabatier and Jenkins-Smith (1993) and Birner and Wittmer (2003).

FIGURE 7.1 Conceptual framework for analysing agricultural policy processes

developed by Hajer (1995, 2006) were used. The main approach was to identify recurring patterns of speech that can be characterised as 'story-lines' (Hajer 2006: 69) which reflect the ways that different actors perceive and explain the phenomena of interest. Combining this approach with the ACF described above, the interviews were also analysed with a view to identifying the strategies used in the creation and deployment of political capital to influence the reform process.

Emerging findings were periodically shared with policy actors and their feedback helped to validate and refine the analysis. The findings were also validated by sharing them in meetings and workshops and by discussing them with peers. Some of the respondents (29 out of 56) were also revisited for purposes of cross-checking and to validate interpretation of data.

The first author was a staff member of MAAIF. Therefore, special emphasis was placed on identifying and addressing potential problems of bias. To uphold confidentiality, the findings from both participant observation and interviews have been anonymised. Each respondent was randomly assigned a number, and they are referred to here as R1, R2, R3, etc.

Creating and using political capital

Here we focus on the strategies that the two coalitions used to influence the design and implementation of the NAADS reform programme. We show how the radical reform coalition was able to shape the NAADS process in line with its policy beliefs. We also explain why the gradual reform coalition had limited influence on the design of NAADS, but subsequently gained increasing influence during the implementation period, to the point of actually succeeding in reversing the reform.

An active radical reform coalition

As indicated earlier, the radical reform coalition was mainly comprised of personnel from the Ministry of Finance, and donors led by the World Bank. This coalition used at least six strategies to influence the reform process.

Use of social networks and public media

This coalition was able to build and use strong social networks, both at national and international levels. Interviews revealed that members of the coalition took advantage of their extensive contacts with international financial institutions to secure the necessary technical assistance to develop the Poverty Eradication Action Plan (PEAP) as the government's overall strategic development framework which was adopted in 1997 (R69, R70). All development programmes were then expected to be PEAP compliant. Significantly, what were to become the core tenets of the Neuchâtel Group and the NAADS reform were already highlighted in the PEAP. These included among others market-oriented agriculture, private sector provision of services, and civil service reforms. The NAADS programme was therefore designed from the outset to be fully PEAP compliant. The link between PEAP and NAADS is an example of the use of international experts to influence local policy making, a phenomenon that is well documented in literature (see e.g. Sabatier and Jenkins-Smith 1993).

At national level, the radical reform coalition was able to win the support of the offices of the Permanent Secretary and the Minister at MAAIF (R11, R14). These offices play a critical role in the policy process for the agricultural sector, and their support was important in making the case that MAAIF backed the reform process. Thus the radical reform coalition was able to create political capital by lobbying and persuading key administrative and policy actors. This is seen in the favourable views of one of the Ministers of Agriculture at the time about the reform. According to his own account, he subscribed to NAADS for a number of reasons. First, the existing policies and structures favoured primary production and yet the new policy was to move towards market-oriented agriculture and value chain development. Second, he engaged consultants to undertake a functional analysis of MAAIF and was convinced by their proposals. Third, the Minister made a trip to South America, which was organised by the World Bank, where he observed this model in practice. According to one of the World Bank officials, the Minister was impressed by the experience in those countries he visited (R67). This all added to the coalition's ability to influence the reform process. Social networks were also used to accumulate information about the reform model that could be used to successfully counter possible policy alternatives.

Furthermore, results from interviews and participant observation revealed that the radical reform coalition adopted a strategy of intensively using mass media, lobbying and advocacy to ensure the programme was favourably received by both civil servants and by the general public. An internet search on the websites of the two leading local newspapers, the *New Vision* and *Daily Monitor* revealed that on average, there was an article on NAADS, whether solicited or unsolicited, every month in one of the papers between January 2004 and December 2008. One member of the gradual reform coalition noted, 'Throughout my 36 years of service, I have not witnessed an agricultural programme that has devoted so much time and resources in the media, as NAADS' (R30). On the whole, the radical reform coalition was able to apply social networks, international human resource and public media to create the political capital necessary to advance the reform process.

Heavy deployment of financial resources

The importance of financial resources for influencing the policy process is well documented (Sabatier and Pelkey 1987; Sabatier and Jenkins-Smith 1993). Document reviews showed that the International Financial Institutions including the World Bank, International Fund for Agricultural Development (IFAD) and other donor agencies financed PEAP and Plan for Modernisation Agriculture (PMA) processes, and later the reform of agricultural extension services (PMA 2000). Interviews revealed that in relation to the extension reform process these institutions commissioned studies and developed working papers; undertook technical and economic analyses; and organised media campaigns to build public support and advertise their policy position (R7, R11, R67). Ultimately they also provided close to 80 per cent of the financial support required for the NAADS programme (Benin *et al.* 2012). In order to qualify for this financial support, government provided assurances that: (i) appropriate measures would be taken to ensure timely processing and enactment of the NAADS Act; (ii) the interim NAADS Secretariat would be formally established not later than the end of the year 2000; (iii) further recruitment of public field level agricultural extensionists would be halted and no further government resources made available to fund such recruitment; and (iv) the NAADS proposal would be fitted within the overall government expenditure framework and government would make available sufficient funds to cover its commitments to the programme. These assurances were incorporated in a letter of Development Policy to Donors written in November 2000 (MAAIF 2000). Accordingly, the NAADS Act 2001 was expedited and approved by the newly elected Parliament.

The financial resources available to the radical reform coalition were used to sensitise and mobilise the Members of Parliament, political appointees and the public through lesson-learning trips abroad, workshops and seminars. This enabled the coalition to influence and gain the necessary support from the key opinion policy and decision makers within government (R24, R65, R68). Overall, the development partners used their financial resources to condition government to embrace wider public sector reforms, in the same way that development assistance internationally had been tied to the public sector reform processes (Asiedu 2003; van de Walle 2007). For most of the period of agricultural extension reform in Uganda (i.e. 2001 and 2010), development partners were contributing an average of US$ 760 million annually to the country as budget support, with the highest contribution of US$ 1.2 billion being made in the 2006/2007 financial year.[1] This represented between 30 and 40 per cent of the national budget. This large contribution to the budget of the Government of Uganda gave the members of the Radical Reform Coalition an important edge in influencing the reform process.

Use of empirical evidence

The radical reform coalition used its financial resources to build the necessary scientific evidence to justify and facilitate the reform process. For example, they hired consultants to draft the NAADS working papers[2] as part of the pre-appraisal document that formed the basis for the design of the NAADS programme. According to interviewees these papers and the evidence that they presented played an important role in the coalition's lobbying of the Minister of Agriculture (R65, R66). Sabatier and Jenkins-Smith (1993) note that scientific evidence may be manipulated and used rhetorically by an advocacy coalition to skew a policy process in its favour.

Interviews with World Bank officials revealed that the Bank established a Thematic Working Group to study agricultural extension reform using the Agricultural Knowledge and Information System (AKIS) framework. This Thematic Group consisted of 180 members from within the Bank and 40 members from outside the bank and it commissioned studies and carried out consultations with, for example, international NGOs and the Food and Agricultural Organization (FAO). The group developed criteria for future World Bank funding of agricultural extension projects, which, perhaps not surprisingly included: empowerment of farmers; private sector involvement; contracting of service provision; monitoring and evaluation; and use of technologies such as those developed by NARO and international research institutions. All projects were to be evaluated based on these five elements of success (R68). As already indicated, these same elements appeared in PEAP and PMA, from which the NAADS programme was developed (MAAIF 2000). In this way, the radical reform coalition succeeded in establishing the evidential basis for reform of the agricultural extension system in Uganda and of agricultural extension more widely.

As the reform process progressed, a series of technical studies generated evidence that was used to demonstrate that the programme was meeting its intended objectives (Adipala *et al.* 2003; Friis-Hansen *et al.* 2004; IFPRI 2004; Scanagri 2005). Interviews with members of the gradual reform coalition showed that results from these studies were used to maintain political support for the programme and build the case for its expansion (R11, R14). As one member of the radical reform coalition observed: 'The programme received tremendous political pressure to expand after the 2004 midterm evaluation' (R26). That evaluation was very positive, indicating that the NAADS model was indeed working.

Setting up separate institutions outside MAAIF run by pro-reform personnel

Another important strategy the radical reform coalition used was to setup the Plan for Modernising Agriculture (PMA) process outside the control of MAAIF by establishing other coordinating institutions such as the PMA Secretariat and Participatory Poverty Assessment Secretariat under the Ministry of Finance. The rationale given for establishing the PMA Secretariat was that the plan was multi-sectoral and addressed development issues beyond MAAIF's jurisdiction (R51). This technically shifted the locus of decision making from MAAIF to the Ministry of Finance. As a consequence, the Ministry of Finance was able to take centre stage in directing the agricultural extension reform process. In creating the PMA Secretariat the radical reform coalition used its social networks to establish cross-ministry linkages, and in so doing reduced the power and influence of MAAIF.[3] In this way, the coalition recruited many actors outside the agricultural sector into the reform process and also attracted competent leadership that could articulate the vision of the coalition within a variety of policy fora. For example, the PMA process that was chaired by the Ministry of Finance engaged about 13 ministries and agencies, and the steering committee alone drew its membership from twenty-four institutions from government, the private sector, academia, parastatals, donors and civil society (PMA 2000).

The strategy of setting up separate coordinating institutions climaxed with the establishment of the NAADS Secretariat as a semi-autonomous agency of MAAIF. The NAADS Act 2001 provided for a NAADS Management Board reporting directly to the Minister. The Board was constituted of 15 members – eight representing farmers, two from agro-industry, one each from MAAIF, the Ministry of Finance, the National Agricultural Research Organisation (NARO) and the NGOs. The Board also included the Executive Director of NAADS who acted as Secretary to the Board (NAADS Act 2001). MAAIF representation on the NAADS Management Board was through the Permanent Secretary. This representation presented an interesting institutional contradiction. While the NAADS Act of 2001 gave MAAIF, whose technical head is the Permanent Secretary, overall responsibility of the programme, his inclusion on the NAADS Board meant that for Board purposes he – and thus MAAIF – was subordinate to the Board Chairman and Secretary. Interviews revealed that the majority of members of both the NAADS Board and NAADS Secretariat supported the policy beliefs of the radical reform coalition (R11, R14, R24, R26, R27).

Keeping MAAIF in a state of transition

The Ministry of Agriculture remained in a state of transition throughout the period of extension reform. Supported by the wider public sector reforms, the radical reform coalition appears to have kept pressure on MAAIF to restructure (R11). Documentary evidence shows that the development partners commissioned studies that recommended a radical overhaul of MAAIF (MAAIF 2002), a recommendation that was resisted by MAAIF. The recommended changes included moving away from the previous departmentalised model to a management structure based on 'corporate teams'.[4] Although the radical reform coalition did not succeed on this front, the pressure it applied kept MAAIF in a state of flux and weakened its ability to provide leadership to the agricultural sector (R1, R5). As MAAIF was a main actor in the gradual reform coalition, this situation significantly weakened this coalition's capacity to act and to influence.

Using the NAADS Phase II design

Funding for the first phase of NAADS came to an end in 2008 and the design of the second phase started in the same year (World Bank 2010a, 2010b). One of the key design strategies for phase II, commonly known as Agricultural Technology and Agribusiness Advisory Services (ATAAS), was to bring together the funding of research and extension. The justification was this would help bridge the gap between research and extension, which was generally considered to be large and problematic. From a political perspective this move brought NARO, a traditional ally of the gradual reform coalition, into the reform fold. This further weakened the gradual reform coalition.

A reactive gradual reform coalition

The gradual reform coalition was comprised mainly of domestic actors from the Ministry of Agriculture, academia, local government and farmer organisations. Evidence suggests that this coalition did not have a strategy to influence events – rather it was largely reactive to the reform process.

Resistance and boycotts

Interviews and documentary evidence reveals that there was some resistance to the reforms, which was manifested as deliberate bureaucratic delays in decision making and implementation, as well as the boycotting of meetings by key actors (R1, R5, R14). According to interviewees, some of the decisions that were critical for the reform's success were either delayed or challenged outright by officials in MAAIF. One example is the conversion of MAAIF extension staff from ministry to NAADS employees. One member of the radical reform coalition stated that 'MAAIF has frustrated efforts to reform and only keeps blaming and opposing any reform efforts' (R53). An example of resistance through the boycotting of meetings was during the mid-term review of NAADS. Out of the 200 participants who attended the meeting, only the Permanent Secretary and the Commissioner in charge of Agricultural Planning and Development represented MAAIF – none of the technical heads of MAAIF departments were present (Scanagri Ltd 2005).

The expiry of the Poverty Eradication Action Plan

The Poverty Eradication Action Plan (PEAP), which served as the overall government planning framework and from which PMA was developed, came to an end in 2008. The government then embarked on the development of the five-year National Development Plan (NDP). The gradual reform coalition exploited this opportunity by initiating the Development Strategy and Investment Plan (DSIP: 2010–2015) within the framework of NDP. Interviews revealed that the strategy was to have a policy document that could replace the PMA, over which the gradual reform coalition had lost control (R1). However, given the policy and institutional structures in place, the radical reform coalition could not be excluded from the DSIP process. As one member of the gradual reform coalition observed: 'Our original idea was to design the plan based on the subsector approach in line with the MAAIF institutional structure that is subsector-based. The donors and Ministry of Finance opted for a functional approach to the design that is not in tandem with the structural arrangements. This will make the plan difficult to interpret by technocrats within a subsector structural arrangement and in turn will complicate implementation' (R14). While MAAIF succeeded in taking control of the DSIP process, it had limited success in integrating the policy beliefs of the gradual reform coalition.

Exploiting the weaknesses of NAADS

With time, it became evident that NAADS was not achieving the desired results. The national agricultural production statistics had consistently shown a downward trend, the number of farmers receiving extension services had not improved (Uganda Bureau of Statistics 2008), and household incomes showed no significant improvement (Benin *et al.* 2012). The NAADS strategy of only supporting farmers who were members of a group and undertaking specific selected enterprises left many farmers without access to extension services. This, together with other design and implementation challenges, undermined the performance of the programme (R5, R11, R14). The National Household Survey (Uganda Bureau of Statistics 2005), for example, showed that only about 10 per cent of farmers received extension services. The gradual reform coalition seized the opportunity to expose what it perceived as the weaknesses of the NAADS design (R11, R14). A case in point is a MAAIF instituted 'Probe Committee' set up to investigate mismanagement of NAADS at the local level. The committee unearthed gross mismanagement at sub-county level, particularly in the procurement of goods and services (R17). Even though the findings were never published, the initiative gave more weight to the gradual reform coalition within the reform process. As NAADS expanded, the private service providers were spread thinly on the ground and this opened the door to inexperienced and unqualified service providers to enter the market. As one of the radical reform members pointed out: 'The failure to lay off public extension staff denied the program the necessary competent service providers in the private sector. This resulted in poor service provision by unqualified private service providers' (R26). As a result, there were widespread concerns about poor service provision and a clamour from farmers to use the MAAIF extension staff who had been side-lined during much of the reform process.

Exploiting the changing of the guard

The presidential and parliamentary elections in March 2006 brought with them new thinking about the agricultural extension reforms. New Ministers of MAAIF were appointed in 2006, 2009 and 2011. Interviews revealed that, unlike their predecessors, these new ministers shared most of the policy beliefs held by the gradual reform coalition (R9, R23, R45). The coalition utilised this opportunity to create political capital and influence the policy process. The failure of NAADS to meet the expectations of beneficiaries allowed the gradual reform coalition to push the case that the programme model was not delivering. Documentary evidence also shows that there were growing governance challenges within NAADS (MAAIF 2009). The financial resources which hitherto had been central for the radical reform coalition to influence the reform process appeared to have lost their 'power' to determine the course of events. This loss was paralleled by formation of political capital by the gradual reform coalition which increased its influence on the reform process. By 2014, the reform process had been effectively reversed, with MAAIF again taking over the agricultural extension services.

Discussion and conclusions

In this chapter we have demonstrated that the gradual reform coalition, dominated by technical staff in MAAIF, agricultural professionals and academia, was largely excluded from the design of the NAADS programme. As a result, the reform programme did not benefit from a significant body of local technical expertise at the design stage.

Furthermore, because of the multidisciplinary nature of agriculture, the omission of MAAIF technical staff from the PMA and NAADS committees, where extension reforms were extensively deliberated, was a serious mistake. A case in point was the assumption by those involved in the design of NAADS that frontline public extension staff were only involved in extension services (MAAIF 2000). In fact, MAAIF extension staff were also involved in the provision of regulatory and disease control functions, which were subsequently neglected, and which affected the outcome of services delivery by NAADS.

The value of resources in influencing outcomes of a policy process was well recognised by Sabatier and Jenkins-Smith (1993). The analysis in this chapter illustrates how power relations between the different actors in the reform process were tilted towards those perceived to be 'compliant' with the reforms, by the nature of the incentive mechanisms integrated in the reform process. The lack of consensus during the agricultural extension reform process meant that the proponents of the reform were able to rely more on their financial resources to get things done than on inclusive processes and widely shared values. Ultimately this significantly weakened the reform process. The donors' large financial contributions to the government's budget, and to NAADS in particular, gave the members of the radical reform coalition a level of influence that the domestic actors could only match once the NAADS programme had run into serious problems. This probably explains the largely passive resistance to the reform process by members of the gradual reform coalition.

One important development that occurred during the design of the NAADS programme was the dissolution of the Directorate of Agricultural Extension Services in MAAIF. This occurred before an appropriate substitute institution could be established. As a result, the technical leadership of agricultural extension was disbanded and did not participate in the design of the NAADS programme. This was probably essential to overcome resistance to reform. However, as we have demonstrated, the quick establishment of the new institutional arrangements under NAADS, in combination with the erosion of existing expertise, undermined the institutional coherence that would have enabled the reform process (Murrell 1992).

The dominant thinking among development partners was that a successful reform process was dependent on the commitment of the Ministry of Finance. Public backing was of lesser importance because agricultural extension is not tangible and its outputs cannot be seen by ordinary citizens (other than farmers). This makes it unattractive for investment at policy level. Convincing Ministers of Finance to allocate sufficient resources to agricultural was a difficult undertaking. The cooperation of the Ministry of Finance in this reform process was a great achievement and presented a good opportunity. However, as highlighted above, ignoring the Ministry of Agriculture undermined this achievement.

We have shown that the NAADS programme was implemented much faster than expected, driven by the desire of the reformers and political pressure to demonstrate success. Many studies, including the World Bank supervision missions, questioned the pace of the programme expansion arguing that it was not matched by institutional capacity (World Bank 2010b). Radical reformers usually pursue speed as a means of destroying the old, on the assumption that the old has no value (Murrell 1992). Our analysis suggests that speed was adopted as a strategy for NAADS to project itself nationally as a viable replacement of the old public extension system. However, the rapid pace of implementation did not make provision for undesirable effects that could emerge from the reforms, such as mismanagement of funds and sub-standard service providers.

Murrell (1992) defined a radical reform programme in terms of its intended end point. He argued that the order for implementation of the parts is determined by the feasibility as much as by strategy, since the emphasis is on doing as much as possible, as soon as possible. This would appear to be the case with NAADS, where the strategies and resources employed by the radical reform coalition were based on a vision of an end-state that lay far in the future, and did not provide room for a pragmatic assessment of present requirements. This pursuit of the end-state, which placed much emphasis on the destruction of the old agricultural extension system, denied the reform process the benefits of policy-oriented learning. Sabatier and Jenkins-Smith (1993) emphasised the need for enduring alterations of thought or behavioural intentions, which result from experience, and which are concerned with the attainment or revision of precepts of the belief system of individuals or coalitions. Policy-oriented learning, therefore, is primarily concerned with changes over time in the distribution of beliefs within coalitions or within the broader policy subsystem. The radical reform approach, as exemplified by the NAADS programme, was undertaken on a wide scale, and yet with limited knowledge of how the policy actors might respond to changes and how the processes worked.

How do we account for the reported corruption, the poor quality of services, elite capture, political patronage and the eventual reversal of the reform (Joughin and Kjaer 2010; World Bank 2010b; Kjaer and Joughin 2012)? Our analysis suggests that the lack of policy-oriented learning, the absence of regulatory mechanisms (Adipala *et al.* 2003), and the disruption of institutional relationships accompanied by declining commitment (ITAD 2008; World Bank 2010b) must be key components of this explanation. As such they helped transform what was originally conceived of as a solution to the ailing public extension system into the problem.

Our analysis demonstrates that policy reform is a deeply political process and is characterised by polarisation of actors who employ diverse strategies to achieve their goals. One of the major strategies that influenced the outcome of the agricultural extension reform in Uganda was the use of financial and human resources. Yet as we have shown, over-reliance on such resources has its own limitations. Consideration of values and processes within the local context is important for long-term sustainability of this kind of reform process.

On the whole, this chapter demonstrates how ideological contestations in policy processes can influence institutional power relations and determine the extent of use of agronomic knowledge and technology. The imposition of market-oriented 'neoliberal' reform legitimated in public discourse through appeals to scientific expertise and narrowly framed technical issues over broader issues related to agricultural extension transformation, reduced the capacity of the Ministry of Agriculture to create and disseminate agronomic knowledge that farmers actually need.

Acknowledgements

The authors wish to acknowledge the very valuable comments received from Anne Mette, Solveig Daneilson, and Joseph Mpagi. They also wish to express their gratitude for the financial support received from the Ministry of Agriculture, Animal Industry and Fisheries of the Republic of Uganda, the United States Agency for International Development and the International Food Policy Research Institute. All opinions expressed in this article are those of the authors and should not be attributed to any of these organisations.

Notes

1   Ministerial Policy Statements of Ministry of Finance Planning and Economic Development from 2001 to 2007.
2   Consultants who produced NAADS working papers I to X (MAAIF and Ministry of Finance 2000: NAADS Programme, Working Papers to the Master Document of the NAADS Task Force and Joint Donor Mission).
3   See the different institutions that constituted the PMA Steering Committee in the PMA Document (PMA 2000: 164).
4   MAAIF (2002): *A Functional Analysis Report of the Ministry of Agriculture, Animal Industry and Fisheries*. Entebbe, Uganda.

8

SWEET 'SUCCESS'

Contesting biofortification strategies to address malnutrition in Tanzania

Sheila Rao and Chris Huggins

Introduction

Recent efforts to promote orange-fleshed sweet potato (OFSP) across sub-Saharan Africa demonstrate the strength of the idea that the production and marketing of new crop varieties can provide health and other benefits to farm families and consumers. Campaigns promoting biofortified crops are targeting millions of farmers in fifteen Sub-Saharan African countries in an effort to address malnutrition (Harvest Plus 2014). Although biofortification has been relatively uncontested as a form of nutritional intervention (Brooks and Johnson-Beebout 2012), and there is some evidence that biofortified crops can positively affect nutritional status (Hotz *et al.* 2012; Ruel and Alderman 2013), it has become increasingly clear in Tanzania that OFSP production and sale by smallholders has not yet reached the scale that was originally envisaged. This is despite a widespread perception that 'Tanzania appeared to have high potential for uptake of the crop' (Waized *et al.* 2015: 3).

This chapter critically examines the ways in which the production of a nutritious crop promoted for sale and household consumption by smallholder farmers, a crop that is well suited to local conditions, and for which a ready market is said to exist, actually relies on complex and highly interventionist approaches. These interventionist approaches involve sustained funding from private donors, training, and related actions from the state, NGOs and research institutes associated with development-oriented agronomy.

We start by briefly discussing recent interest in 'nutrition in agriculture' in sub-Saharan Africa, and then situate the promotion of biofortified sweet potato within this discourse. We then examine the benefit claims made by organizations promoting OFSP production and consumption in Tanzania, such as empowerment of women, improved income for farming households and improved child nutrition. Drawing from recent ethnographic fieldwork conducted by the first author,[1] and an extensive literature review, we demonstrate that the approaches used by various research, development and private sector actors to promote OFSP involve the commercialization of the crop at a large scale, which is likely to compromise the claimed benefits or have unintended negative consequences for women. Specifically, we highlight how efforts to commercialize the multiplication of OFSP vines, as the latest strategy to take OFSP 'to scale', offers financial benefits to farmers, yet often favours male farmers.

We argue that despite promises of rapid, cost-effective uptake and claims about innovation, efforts to achieve 'multiple wins' through rapid adoption of OFSP are negatively affected by social, cultural, gender and economic constraints that are common to many agricultural development efforts in East Africa.

Undernutrition and biofortification

Malnutrition remains a widespread problem in sub-Saharan Africa (SSA). Chronic malnutrition – stunting, or low height for age – affects 40 per cent of children under five in SSA (UNICEF 2013). To take another measure, 43 million children in SSA under the age of five are vitamin A deficient (Stathers *et al.* 2015). In Tanzania, 35 per cent of the under-fives are considered stunted and 11.5 per cent severely stunted (UNICEF 2013; IFPRI 2014). The World Health Organization (WHO) considers these levels to be 'very high' (United Republic of Tanzania 2014). About 72 per cent of children aged 6-59 months receive vitamin A supplements, against a Tanzanian government target of 80 per cent (Stathers *et al.* 2015).

As one of the first countries to join the Scaling Up Nutrition (SUN) movement in 2011, the government of Tanzania has made commitments to address malnutrition, and there is some evidence that progress is being made (UNICEF 2015). Efforts to address malnutrition in Tanzania are heavily influenced by global thinking and regional initiatives. Over the past decade or so, there has been much emphasis on the creation of stronger linkages between agriculture and nutrition, as evidenced by the L'Aquila Joint Statement on Global Food Security, for example, which calls for a 'cross-cutting' approach to agriculture, food security and nutrition (G8 2009). Other recent initiatives emphasize the role of markets in meeting the nutritional needs of citizens, through advances in agronomy, business and food science (Pittore and Robinson 2015).

The micronutrient approach and biofortification

In the 1990s, the focus in influential nutrition circles shifted toward micronutrients. The concept of 'hidden hunger' emerged partly because of new scientific arguments that deficiencies of micronutrients, including vitamin A, iodine, zinc and iron, could be responsible for stunting and other biological developmental problems (Brooks and Johnson-Beebout; 2012; Haddad 2013; Kimura 2013). The three main responses to hidden hunger have been supplementation, fortification (addition of nutrients to commodities during processing or manufacture, such as the addition of iodine to salt), and biofortification (increasing the nutrient composition of crops through genetic manipulation).

Earlier interventions to address vitamin A deficiency focused on supplementation, and were largely state-dependent and targeted mostly at young children. In Tanzania, two campaigns introduced by Helen Keller International provided free supplements to hospitals across the country. However, coverage was not complete, and while supplementation programmes such as these were shown to be effective in lowering child mortality, they did not improve the long-term health of children (UNICEF 2007). Recently, due to a desire to reach nutrition goals more rapidly and cost-effectively (McDermott et al. 2013), food–based solutions, including biofortification, have become the preferred route. Biofortified sweet potato is the first crop for which there is evidence of the positive effects of biofortification on nutritional outcomes (Saltzman et al. 2013).

The experiences with 'golden rice', which is high in Vitamin A, and quality protein maize (QPM) paved the way for further investments in biofortification research and development (Brooks 2013). This enthusiasm prevails despite serious delays in dissemination of the materials and controversial and often inconsistent evidence of nutritional benefits from initial feeding trials (Brooks and Johnson-Beebout 2012). Orange-fleshed sweet potato (OFSP), bred 20 years ago to accumulate higher vitamin A than traditional varieties (which tend to be white or yellow in colour) has been described as 'a flagship of biofortification efforts' (Waized et al. 2015).

There have been a number of initiatives – some ongoing – to support research (into breeding, vine multiplication, processing and product development), promotion, advocacy, fundraising and overall awareness-raising around OFSP. Vitamin A in Africa (VITAA) was a consortium of over 70 organizations and representatives from several SSA countries which operated from 2001 to 2006 and helped boost interest in OFSP. The ten-year Sweet Potato for Profit and Health Partnership (SPHI) is currently promoting OFSP in 17 African countries (CIP 2011, cited in Waized et al. 2015). The two phases of a project called Sweet Potato Action for Security and Health in Africa were implemented through SPHI by The International Potato Center (CIP), where the first phase focused on new variety development and the second on vine distribution and markets (Waized et al. 2015). CIP has been heavily involved in the promotion of OFSP in Tanzania, and across Africa, including the Reaching Agents of Change project, which aimed to increase 'investment' in OFSP and build capacity of key actors to plan OFSP-related programmes (Mbabu et al. 2015).

The CGIAR has been deeply engaged in development and promotion of OFSP, because sweet potato is a staple crop in certain regions of SSA. However, there are differences in approach between the two parts of the CGIAR that are most closely engaged. CIP promotes biofortified sweet potato as part of a broader portfolio of locally bred sweet potato varieties (including non-biofortified types) developed by national breeding programmes. CIP's efforts in Rwanda also include the production and marketing of 'golden biscuits' and other processed food products made from OFSP. Harvest Plus, on the other hand, puts the technology of biofortification at centre stage, and promotes OFSP as part of a suite of biofortified staple crops. One biofortified crop is promoted per country: in Rwanda, Harvest Plus has promoted iron-beans, for example, while it has focused on OFSP in Uganda since 2004. Support for OFSP initially involved CIP and Harvest Plus working together, but has evolved more recently into separate initiatives in different countries. The CGIAR's efforts in this domain are heavily dependent on funding from private foundations such as the Bill and Melinda Gates Foundation (BMGF). Thus, while they share similar objectives toward OFSP, and share funding sources, there is evidence of growing compartmentalization which reflects the diversity of projects developed through, for example, CGIAR centres, NGOs, national research centres and businesses which will be further discussed below.

Whether the starting point is the crop or the idea of biofortification, the implicit premise of much work on biofortification is that 'breakthrough science' has the power to solve 'complex social problems' (Brooks 2010: 14). Complex realities are simplified, first, into a particular form and then into an 'isolable problem', for which technical solutions can be identified (Brooks 2010: 17, citing Anderson *et al.* 1991). Part of the justification for biofortification is cost-effectiveness: 'investments in micronutrients have higher returns than those from investments in trade liberalization, in malaria, or in water and sanitation … No other technology offers as large an opportunity to improve lives at such low cost and in such a short time' (Behrman *et al.* 2004).

An important part of the attraction of OFSP is that sweet potato is widely considered a 'women's crop'. As women are portrayed as the guardians of children's health and nutrition, and household food security more generally, even though evidence remains inconclusive (van den Bold *et al.* 2013; Gilligan *et al.* 2014), it is assumed that women who adopt OFSP will feed it to their children (and eat it themselves), resulting in nutritional benefits. As discussed below, this simple and compelling logic seldom plays out so neatly in programmes to promote OFSP production and use. Projects rarely address the underlying gender relations and inequalities (including labour, time and resource allocation) that prevent both men and women from benefiting from these kinds of externally supported initiatives for farmers.

While the hidden hunger agenda was driven in part by technological advances, it was also embedded within broader processes including the neoliberal commodification of science. Emphasis was on market-ready technologies, which could open up new markets and generate profits, at a point when structural adjustment programmes had forced the state to roll back its involvement in the agricultural sector (Kimura 2013). Micronutrient 'solutions', including crop biofortification, were thus a test case for the new model of public–private partnerships (Kimura 2013) and as such, can be seen as part of the broader neoliberal transformation of agronomy (Sumberg, Thompson and Woodhouse 2012). Biofortification and the associated OFSP narrative fit well with the now dominant, market-oriented paradigm in smallholder agricultural development (McDermott *et al.* 2013). The CGIAR and international NGO-led initiatives, largely funded by BMGF, assumed that markets for both planting material (vines) and for roots would develop rapidly, without the need for sustained external intervention.

'Win–Win' narratives

The potential of various biofortified crops has been heralded with some spectacular claims (Cullather 2004; Brooks 2010; Brooks and Johnson-Beebout 2012). This reflects in part a tendency for those promoting 'Green Revolution for Africa' to argue that with technology, agricultural transformation can be rapidly achieved while avoiding the problematic aspects of the Asian Green Revolution (Conway 1999; Anan 2008; AGRA 2014). But more broadly, this kind of hype is rife within international development (Mosse 2005), which increasingly uses the 'success story' to reach both specialists and non-specialists (Sumberg, Irving *et al.* 2012). Hunsberger (2010) uses Tsing's (2000) concept of 'the economy of spectacle' to explore how NGOs in Kenya replicated a narrative of success in jatropha production, despite the limitations and problems experienced in making jatropha profitable for smallholders. She argues that Kenyan NGOs were 'being encouraged by donors' excessive focus on "success" stories as a condition for future funding, with the result that NGOs may cover up their failures' (Hunsberger 2010: 227). NGOs are limited in what they can report for fear of losing funding or the possibility of future engagement in a particular development agenda (Green 2012).

The notion of spectacle is also of relevance to OFSP. Sumberg, Irving *et al.* (2012) note that success stories intended for policy-makers and non-specialists are often packaged along with captivating images. Some of the OFSP promotional and educational materials use images and paraphernalia including orange t-shirts, and orange-painted cars and trucks. Videos, songs and other products are also used. Such strategies can be characterized as '360-degree' marketing (Igoe 2010: 377). Images of women smallholder farmers are central to much OFSP promotional and policy literature, reinforcing the importance of women's role in growing and selling the crop. Research conducted by the CGIAR and national research centres feeds into broader development objectives that emphasize commercial benefits and nutritional gains for both men and women.

The 'win–win' narratives that are used by major OFSP programmes are partly a response to the lack of interest in initiatives which only target vitamin A deficiency (Mbabu *et al.* 2015). In response, these narratives suggest that increased planting of OFSP can result in:

- improved nutrition for children and household members more generally
- improved household income through sales
- improved welfare for women through their involvement in producing and marketing OFSP

- positive impacts for poor farmers, as OFSP can be grown on marginal land with minimal agrochemical inputs.

(DFID 2015; HKI 2012)

For example, one brochure states that OFSP can 'cost-effectively improve nutrition, empower women, and increase income earning opportunities, even for the poorest household' (RAC 2012: 1), while another promotes 'biofortification to improve nutrition, health and income' (Mbabu *et al.* 2015: 16). It is also suggested 'OFSP can be used as an entry point for changing behaviours' (RAC 2012: 1). These claims of multiple wins are representative of other development-oriented agronomy agendas influenced by investors in humanitarian and international development work, such as BMGF.

The claims, and the ways in which they are framed, also reflect the influence of New Public Management on agricultural research and development (Andersson and Sumberg 2017). These include elaborating targets, identifying quick wins, schedules, theorizing impact pathways, and demonstrating 'value for money'. Much emphasis is placed on 'investment' in OFSP (Mbabu *et al.* 2015), in ways reminiscent of the 'business case' model used by the UK Department for International Development (DFID). It is important to note that the alliances of organizations promoting OFSP elicit investment from donors and the private sector, implicitly locating the state as one among many investors, rather than the key one. In particular, private foundations and bilateral funding arrangements (BMGF 2013) are deeply involved in creating and investing in such agendas, that directly respond to specific development goals such as nutrition security, or empowering women; however, the pressure for 'quick wins' leads to less emphasis on contextual differences, complex gender relations and local level governance. Rather the emphasis is on demonstrating short-term economic gains, whether at the household or country-level (Andersson and Sumberg 2017).

A brief history of OFSP promotion

Over the last 20 years, the promotion of biofortified or orange-fleshed sweet potato has expanded across SSA, and OFSP is projected as one of the most successful cases of crop biofortification (Saltzman *et al.* 2013). The first known study of OFSP, in South Africa, showed an increase in liver vitamin A stores in 4–8 year-old schoolchildren (van Jaarsveld *et al.* 2005).[2] These initial results encouraged further investments in studies of the efficacy of OFSP as a source of vitamin A for children.

One of the earliest promotional programmes took place in Mozambique starting in 2002 and involved a number of partners. Different dissemination strategies were used, and public awareness and marketing activities sought to integrate OFSP into existing sweet potato cropping systems (Low *et al.* 2013). OFSP vines were initially distributed free of charge to women farmers with young children. Women were also given opportunities to learn cultivation techniques and ways of dealing with pests and diseases. Marketing was also supported through the establishment of orange-painted kiosks in local markets (Low *et al.* 2013). Radio programmes were used to reinforce key messages around nutrition and marketing. These first initiatives showed potential, but ultimately the varieties performed poorly.

CIP's work showed that through campaigns targeting selected communities with information and free planting material, OFSP decreased vitamin A deficiency. However, these strategies were seen as expensive, at a cost of US$ 79 per beneficiary. Harvest Plus conducted a follow-up study in Mozambique and Uganda, which aimed to improve cost-effectiveness through the utilization of existing groups rather than engaging in group formation (Low *et al.* 2013). Results showed high rates of adoption (65 per cent) in project households (compared to 4 per cent in control communities) and an increase in land investment for OFSP production (Harvest Plus 2012). These projects reduced costs to US$ 52 per beneficiary in Mozambique and US $56 in Uganda (ibid).

The Mozambique project evaluation recommended that trained farmer-multipliers (who have ready access to a water source) be supplied with OFSP planting materials, so that they can then distribute vines of the improved varieties in their local community. This strategy was preferred to mass distribution efforts (such as a voucher system) (Low *et al.* 2013). This example led to recommendations for larger investments into national breeding programmes, to produce more OFSP vines and enable 'scaling up' from locally bred varieties. However as will be described below, vine multiplication efforts complicated the win–win claims of OFSP promoters.

OFSP promotion in Tanzania

The initial trials and action research projects in Uganda and Mozambique led to an expansion of projects and initiatives in other sub-Saharan countries, including Tanzania, where sweet potato is the third most important food crop. Sweet potatoes can be grown from sea level to an altitude of 2350 m. With a short growing season of four months, and some drought-resistance, sweet potato is considered a household-level food security crop (Mmasa *et al.* 2014). Women and men commonly share field preparation activities (including ploughing, ridging, and preparing beds) and men typically do some weeding, but much of the other work is done by women (vine cutting, weeding, harvesting, bagging, transporting to market and selling). Marketing of traditional varieties of sweet potato is mostly informal and small-scale (Mmasa *et al.* 2014). Because of their role in selling OFSP, women are assumed to have control over how profits will be spent after sales (Mmasa *et al.* 2014), however, further research is required to examine more context-specific aspects of decision making and labour relations at the household and village levels (Sindi and Wambugu 2012).

There are three main areas in Tanzania where sweet potato is consumed regularly: in the regions bordering Lake Victoria, and in the Southern and Coastal regions. The Lake zone, where sweet potato is the second most important crop after cassava (Sindi and Wambugu 2012) was one of the first areas of Tanzania to be targeted for the promotion of the new biofortified varieties. Many of the varieties initially introduced in Tanzania had been developed in Uganda and Kenya; newer varieties bred in Tanzania were introduced later. The areas along the lake, including Mwanza, were thought to have the potential to become major OFSP production zones, due to the regular access to water, two growing seasons and the importance of sweet potato in the diet (Kapinga *et al.* 1995). From 2002 to 2014 a total of ten different OFSP promotional initiatives took place in Tanzania, of which six were located in the Lake zone; they were supported by a number of different donors including private foundations and development agencies.

Early interventions

For clarity we divide large-scale OFSP interventions in Tanzania into a 'first wave' (approximately 2004–2011) and a 'second wave' (from 2012). This allows for a more nuanced account than that provided by Waized *et al.* (2015), who focus on 'pre-farmgate' and 'post-farmgate' pathways for OFSP promotion. While our first- and second-wave model is only approximate, it is based partly on existing schema for promotion of biofortified crops, such as that put forward by Bouis and Islam (2012: 2) who posit three stages:

> The first level of scaling up requires that a critical mass of poor farmers adopt the biofortified crop and feed it to their families … At this level informal diffusion is a pathway by which the food is introduced to others in the community. At the second level, markets for the biofortified crop need be developed to provide farmers with an outlet for marketable surplus, thus reaching nonfarming or rural households that are net buyers of food. This second level is driven by … reaching out to medium-scale producers, and developing local markets and demand for products made from biofortified foods, still largely in rural areas. At the third level, the private sector becomes the main driver of the diffusion process … value chains for biofortified crops can be developed to produce value-added tradable products in order to mainstream biofortification.

In the first wave of interventions emphasis was generally at the farm level, and specifically increased production by targeted households for own consumption. These interventions used varieties produced by international and national agricultural research centres as part of the Sweet Potato Action for Security and Health in Africa (SPHI) initiative. Some varieties came from neighbouring Uganda, and were tested through the national agricultural research centre. These were distributed to farmers free of charge or were heavily subsidized. Similar to the initial promotion activities, initiatives targeted regions where rural farming households regularly consumed sweet potato. Catholic Relief Services (CRS) and Helen Keller International (HKI) along with the Mwanza-based organization,

Tanzanian Home Economics Association (TAHEA), initially distributed planting material in several districts in the Lake zone. They used similar strategies to the CIP and Harvest Plus projects in Mozambique and Uganda: women with children were given free vines to plant, multiply and distribute locally. Vines were also given to existing farmer groups in the Mwanza region, and NGOs conducted technical training activities through demonstration plots and farmer visits. HKI, CRS and TAHEA worked closely with the network of three national agricultural research centres in the targeted regions, which were also multiplying OFSP vines.

Most projects integrated vine multiplication as part of the production process, rather than as a separate or independent activity. This encouraged farmers and farmer groups to continue using cuttings from the original vines distributed to them by the NGOs. One important aspect of OFSP production is that for yields to be sustained, fresh planting materials must be used every few years – otherwise disease can accumulate, resulting in reduced yields (Ogero *et al.* 2015; Schmidt 2014). Sweet potato vines cannot be stored for long periods (Gilligan *et al.* 2014), and therefore, ensuring a supply of clean vines is important to the promotion model. However, the long-term prerequisites for quality vine multiplication and distribution were not comprehensively factored into the first phase of interventions in Tanzania.

Given that OFSP vines were scarce in comparison to white or yellow varieties, some projects provided farmers with both OFSP and traditional varieties, mixed together.[3] Promotion of OFSP by some organizations, and its particular nutritional advantages, was therefore somewhat subordinate to a broader objective of farm-level modernization (through new crop varieties). Alternatively, this strategy can be understood as a way of integrating OFSP into existing systems of sweet potato cultivation and markets. There was a trade-off, therefore, between highlighting the particular qualities of OFSP and achieving greater scale through mixing it with more popular varieties. This trade-off accounts, at least according to one study, for the low numbers of people growing OFSP in different regions in the Lake zone by 2012. Sindi and Wambugu (2012) reported that only 7 per cent of their sample grew OFSP compared to 99 per cent who grew white and yellow varieties. Researchers also noted that few people in the former project areas continue to use the OFSP recipes that had been promoted, and that OFSP is rarely consumed outside of the project areas (Schmidt 2014).

One of the anticipated outcomes of OFSP promotion projects was that local markets for OFSP (both roots and vines) would quickly emerge, given the pre-existing markets for white and yellow sweet potato varieties (Waized *et al.* 2015, citing Temu *et al.* 2014). However, this was not necessarily the case. In one example, some young entrepreneurs produced and sold approximately 2,000 bags of vines; 76 per cent of these were purchased by INGOs or government agencies for further OFSP dissemination activities, with only the remaining 24 per cent being sold directly to farmers (Namanda *et al.* 2006). Local markets were much smaller than expected.

Following such experiences, major donors provided funding for national research institutions in Tanzania (such as the Ukiriguru Research Institute in Mwanza) to develop more OFSP varieties and to increase vine production. Private actors such as Crop Bioscience Limited also stepped in to meet the demand for vines by NGOs and other OFSP implementing partners. A parallel USAID-funded Tanzanian Agricultural Production Program (2011–2015) funded the company's start-up costs.[4] More recently, a second USAID-funded programme, Viable Technologies for Sweet Potato in Africa (VISTA) has focused on increasing the availability of sweet potato planting material in the southern regions of the country.

More recent interventions

The second wave of interventions in Tanzania were more focused on creating sustainable markets for clean planting materials, and for boosting productivity through marketing, processing and the commercialization of products from OFSP. As such, these initiatives shifted the focus to post-farmgate opportunities. However, the importance of planting material remains: since the lack of clean planting material is seen as one of the main limiting factors to scaling up, a priority in the Mwanza region is to increase its availability. The introduction of net tunnels in recent projects aims to increase access to clean material and also decrease risk of disease and virus contamination.

Based on the earlier interventions in Uganda and Mozambique, major organizations such as CIP concluded that mass distribution of vines (either through a voucher or kiosk system) was too complex to replicate, preferring instead a decentralized vine multiplication (DVM) system (Low *et al.* 2013). The DVM approach provides technical support to small- and medium-scale farmers who are capable of multiplying and marketing clean vines. CIP managed efforts to increase vine multiplication in the Mwanza region through their projects *Marando Bora* ('healthy vines') and *Kinga Marando* ('seeds free from disease'). A similar project managed by an international research institute and funded by BMGF multiplies vines in locally sourced and constructed tunnels situated close to the lake, and these are later transplanted into upland fields for growing out.[5]

Farmers are pre-selected for such DVM projects based on their past involvement in OFSP production, and their capacity to manage the nurseries or tunnels (including time and resource availability). Vine multiplication requires a lot of water applied by hand or using other forms of irrigation, so only farmers located adjacent to a water source, or in wet valley bottoms, can become multipliers. While some districts targeted women farmers to engage in DVM, the majority of efforts in Misungwi and Ukerewe districts were led by male farmers who grow other commercial crops, have larger land holdings (often close to a reliable water source) and who have direct contact with district extension officers.[6]

Major organizations have disseminated success stories of women vine multipliers (HKI 2012; CRS Tanzania 2012), reflecting the broader importance of success stories within development-oriented agronomy (Hunsberger 2010; Sumberg,

Irving *et al.* 2012). These women were supported to attend international meetings on OFSP and biofortification, for example in Uganda and Kenya, where they shared their stories. While they fondly remember their travels, few of them currently have regular access to clean planting material or produce OFSP commercially.[7]

Based on recent fieldwork, in the remainder of this section we present two examples of predominantly female farmer groups involved in vine multiplication. Both groups have produced and sold large volumes of vines, although they now find themselves in quite different situations. The first group is in Misungwi District, Mwanza Region. This is an area where both yellow and white sweet potatoes are popular, and where NGOs focused on OFSP promotion. The first example is the original group included in initial sweet potato production in 2004. Over the years, they developed a strong relationship with a local NGO engaged in larger OFSP projects. In 2014, the group received orders for vines after the airing of an advertisement on the radio programme, *Kilimo Chetu* ('Our Agriculture'). Phone numbers of vine suppliers, including this farmers' group, were given out on air. One farmer reported filling two large orders from Iringa and Dodoma districts, areas where OFSP had not been promoted. In late 2015, *Radio Maria* continued to broadcast the phone numbers of suppliers, and farmers in Misungwi occasionally receive orders as a result. With the help of the local NGO that initiated OFSP promotion in the area one farmer was able to shift her business from selling home-brewed alcohol (from cassava) to OFSP. When the radio programmes ended, there was still some interest, but sales became more sporadic. Early rains disrupted a recent planting season and therefore other crops (such as rice and okra) took priority for her household. Another farmer depended more on vine sales, but they were not enough to meet basic household needs. Once the implementing organization stopped their project in the area, sales and income from vine sales declined. Two male group members continued to pursue direct engagement with donors and researchers at the Ukiriguru Research Institute through ongoing tunnel programmes, whereas the female group members receive indirect benefits from the men's activities, or benefit through sporadic NGO engagement. There were only a few cases where women farmers met most of the criteria of the vine multiplication programmes, including access to land in wetland areas, time and labour. This example shows that even in areas with historic demand for sweet potato, there may not yet be sufficient demand for OFSP to make it a commercially successful endeavour.

The second group is located in Hai District (Kilimanjaro Region), where sweet potato has not historically been widely grown. The group was introduced to OFSP through a short segment on local television news in 2012, when the second wave of interventions were being launched, and new varieties released. In this region, OFSP introduction originated from two main sources: a USAID-funded project, Tanzania Agricultural Agribusiness Program (TAAP), and a smaller NGO, funded by a Danish cooperative group. The Chair of the local Imani farmers' group purchased a number of vines from Crop Science Limited through TAAP. Staff from TAAP assisted members of the group with training on vine multiplication and sweet potato production. Through the TAAP network, the group connected to a buyer looking to purchase a large volume of vines. The group members aggregated their vines and sold them successfully. A total of 15 large sacks of vines collected from several group members earned the farmers some additional income. However, based on interviews with the group in 2015, some continue to grow only for home consumption, while others stopped completely due to lack of planting material or lack of consistent market. Another farmer in neighbouring Rundegai district switched one acre of land from maize to sweet potato when she received vines from an NGO. She earned some income the first year (in 2013) but sales and production declined in the second year. In 2015, she decided to grow less OFSP and more maize, since there was no guaranteed market. This example shows the difficulties of sustaining OFSP production in areas where there is no pre-existing demand for sweet potato.

Both these examples suggest that there are possibilities for women and women's groups to benefit from vine multiplication and sales of both vines and roots. However, there are outstanding questions about the sustainability of these activities if or when the external support ceases. As noted by Waized et al. (2015), local markets for OFSP vines and roots are still limited in Tanzania. To purchase vines requires not only that smallholder farmers pay for the vines themselves, but also any transport costs to visit the multipliers if they are not living nearby. The costs of the vines and transport can be unacceptably high, given that commercialization of OFSP is limited and demand for roots remains low in central markets. As a result, there is not enough planting material or commercial OFSP in circulation to meet the scaling out objective of the OFSP promoters. The economics of commercial vine multiplication are also uncertain, as customers will only purchase new vines every three to four seasons. The Imani farmers group sold vines in addition to their other cash crops through established markets. Therefore selling OFSP represented supplemental income that was only realized when there was a buyer. With the farmer group in Misungwi district, sale of OFSP vine was the basis upon which the group was formed, but they have dropped drastically in recent years. OFSP is always mixed with yellow and white varieties, and is only preferred by consumers where it has been previously grown as a household crop.

According to CIP and an NGO field officer who has worked in OFSP promotion since 2004, in preference to purchasing clean vines from multipliers, many smallholders prefer to exchange planting materials with friends, neighbours and other farmers (David et al. 2012). Vines are usually exchanged without any payment; sometimes they may be bartered for other farm produce.[8] Sindi and Wambugu (2012) reported that most respondents borrowed vines from neighbours, even if they purchased a small quantity of vines from others. However, 'strangers' (for example, farmers from distant areas who have been put in touch with multipliers through the radio or by NGOs) may be charged for the vines.[9] Several farmers interviewed in Mwasonga village continue to use vines multiplied from original vines received in 2004, which means that the quality of the material is likely to be low. For crops grown for home consumption the social realm, rather than the market, is an important source of agricultural advice and material. DVMs, voucher systems, targeting poorer households and women growers do not address the underlying inequalities between women's and men's access to financial, social and physical capital (including their access to land, water), labour and time.

Discussion and conclusion

Our analysis suggests that the idea that OFSP can be seamlessly introduced into existing smallholder production systems and local markets is not realistic. As one project stated, 'farmers and marketers were slow to adopt OFSP since the crop was highly perishable, and so the market infrastructure had to be created. Agroprocessing techniques are slow to develop and did not take into account the speed at which the OFSP value chain needed to evolve' (Mbabu *et al.* 2015: 22).

Part of the problem may be the lack of prioritization of nutritious crops such as OFSP within state extension and research services. While the Government of Tanzania has made some commitments to prioritizing nutrition (HKI 2012), the 'Kilimo Kwanza' (Farming First) policy emphasizes economic development through increased productivity and commercialization, primarily at large-scale (Tanzania National Business Council 2009; FAO 2013a; Robinson *et al.* 2014).

In the last few years the Government has committed itself to nutrition goals, particularly through SUN, and has recruited nutrition officers at the district and regional levels (Save the Children Tanzania *et al.* 2012). In 2014, the first ever Public Expenditure Review of the nutrition sector was conducted, as well as an official review of the implementation of the first three years of the National Nutrition Strategy. Between 2010 and 2014, stunting levels for children under five decreased from 42 per cent to 35 per cent (United Republic of Tanzania 2014). However, government efforts still rely heavily on the Ministry of Health (rather than multi-stakeholder initiatives), government disbursement for nutrition activities is often delayed or limited, and OFSP does not feature prominently in Government policies and programmes. Biofortification is mentioned only in passing in the National Nutritional Strategy (United Republic of Tanzania 2011, 2014). Moreover, 'all the decisions on OFSP would be made in the Ministry of Agriculture' (Mbabu *et al.* 2015: 17) pointing to the challenge of coordination across Ministries. The limited capacities of the extension services also play a part. Due to the high ratio of farmers to extension workers, '60%-75% of households in Tanzania are estimated to have no contact with research and extension services' (HKI 2012: 19) and women farmers are particularly unlikely to receive extension advice (United Republic of Tanzania 2006, 2014). Despite the high visibility of OFSP internationally, the Government of Tanzania has not systematically integrated it in its agronomic plans and activities, even in the regions where sweet potato is a staple crop (Waziri 2013). As is common in many other countries (Sumberg, Thompson and Woodhouse 2012) the state no longer plays an overarching role in coordinating agricultural development efforts. As a result, approaches to OFSP promotion are diverse and sometimes contradictory.

But our findings point to more than a lack of government commitment, poor inter-ministerial coordination and an over-burdened extension service. OFSP promotion demonstrates some of the limitations and contradictions within development-oriented agronomy. This chapter is not arguing whether or not OFSP can potentially address malnutrition. The consumption of OFSP at household level has been shown to improve childhood nutrition in Uganda (see e.g. Hotz *et al.* 2012; Gilligan *et al.* 2014), and reduce diarrhoea in children in Mozambique (Jones and de Brauw 2015). However, for maximum benefits to be achieved, households must switch from white or yellow varieties to eating mostly or exclusively OFSP. This is highly unlikely to happen in Tanzania due to the limited supplies of OFSP planting material in key production areas, as well as continued preferences for the yellow and white varieties (HKI 2012).

The claim that OFSP production can increase household income is based on the assumption that a sustainable market for OFSP will develop and that OFSP will command higher prices than white and yellow varieties. The emphasis on markets for nutritious crops is part of a broader agenda of commercialization within development-oriented agronomy that also reflects the changing realities of funding for agricultural research, especially the influence of private foundations. As described above, to date there is little evidence of the development of markets for OFSP 'at scale'. Future market development will require continued nutrition education, as well as functioning and financially sustainable vine multiplication systems and, eventually, heightened demand for processed or manufactured products based on OFSP. All of this points to continued and probably large-scale, yet compartmentalized interventions by the state and other development actors, where nutrition advocates, agronomists, NGOs, media and research institutes compete for funds to pursue various aspects of OFSP work. While intervention in commodity chains is mentioned in some of the promotional literature on OFSP, the need for such large-scale interventions tends to undermine the claims of cost-effectiveness and rapid adoption and scaling up.

Even though recent studies point to significant gender challenges in OFSP production, these have not been addressed either by past projects focused on production, or current projects that emphasize vine multiplication and commercialization. Projects have assumed that because sweet potato is a woman's crop in Tanzania, women would benefit, without interrogating women's roles in, for example, vine multiplication. This limited engagement with gender directly affects the OFSP 'quadruple win' narrative – child nutrition benefits, increases in household income, women's empowerment, and reaching the poorest households. The claim that women will benefit from the sale of OFSP – a claim featured in numerous success stories featuring female farmers – is not necessarily borne out by the available evidence. In some cases, commentators jump from an observation that women provide most of the labour in growing and selling sweet potatoes, to an assertion that they will control the proceeds of the commercialization of OFSP.

In other words, an assertion that women's labour inputs translate into 'bargaining power' over the profits. Meinzen-Dick *et al.* (2011) suggest that integrating nutrition into agricultural development projects may offer opportunities for women to have more bargaining power around decision-making processes. However, this may not necessarily be the case for decision making around income: a study of white and yellow sweet potato value chains in Mwanza, Tanzania found that, 'women … are normally left out in the decision making on the use of income they get after sales' (Waziri 2013: 126). Only 43 per cent of respondents said that women are involved in decision making on the use of money accrued from sweet potato (ibid.) Evidence from Kenya suggests that while women did have some control over profits from OFSP, men had control over household income more generally and hence women needed to inform their husbands about how OFSP income was spent (Hagenimana *et al.* 1999). Gilligan *et al.* (2014: 3) show that 'introduction of biofortified crops may lead to complex changes in gender roles for crop choice decisions and crop production that will be shaped by intra-household bargaining power'. Women's roles in the OFSP commodity chain will also be shaped by the development of vine multiplication systems, which, to date, appear to be dominated by men. The fact that vine multiplication requires good access to water for irrigation, also tends to work against women's involvement, as the most strategically located fields are more likely to be controlled by men.

The idea that OFSP can reach the very poorest relies on one of two pathways: either that the poor will be able to grow OFSP and benefit from consumption of their own produce; or that if they do not grow it they can access it through markets. The first pathway rests on the assumption that poor households will be consistently able to access and afford clean vines (Waziri 2013). Clearly a market for vines of improved varieties is one route, but other studies suggest that social relations remain important in the circulation of vines. In at least some cases the poor prefer to access vines through friends, relatives or neighbours (Sindi and Wambugu 2012). This resistance to the commercialization pathway raises questions about the viability of market-based efforts to promote OFSP. The commercialization pathway depends on increasing the number of poor households who rely on markets to source their food (Robinson *et al.* 2014), with all the uncertainty that this implies.

In conclusion, OFSP has been presented as a technology that can be rapidly and easily adopted by smallholder farmers with significant benefits to both the rural and urban poor. It is often assumed to be more cost-effective than other nutritional interventions. However, the results of 'first wave' OFSP promotion activities demonstrate that individuals in project areas will switch from yellow and white sweet potato to OFSP only when distribution of OFSP vines is accompanied by nutritional awareness-raising activities. Furthermore, it appears that such outreach and awareness-raising activities, and promotion more generally, must be sustained over an extended period in order to have a lasting effect. This conclusion accords with other academics who argue that, 'OFSP could be a really promising option *within a broader strategy* to improve nutrition … However, the vast communication effort and huge amount of funds involved should not lead us to overestimate the real impact achieved so far' (Virchow 2013: 48).

Moreover, we have noted the difficulties involved in ensuring a sustainable supply of clean vines, the gender dynamics of vine multiplication, and the tendency for vines to be freely distributed. These show the complexities of employing market-based strategies to go 'to scale' and reveal tensions among the loosely coordinated strategies of state, international and local actors in promoting OFSP.

The win–win claims of the OFSP investors, promoters, observers and advocates, and suggestions of a smooth integration into household production and consumption systems, and local markets for food, can be interpreted as a means of stimulating interest and investment. To date there is little evidence that such win–win scenarios can be achieved quickly. Indeed, some projects seem to have devoted more effort to disseminating promotional messages than achieving their goals. In a rare example of development organizations admitting their weaknesses, one major OFSP promotion project revealed that 'the team could not provide ready answers to investors on how much to invest and on what components. Where funding was secured, RAC could not provide tangible guidance on how to implement OFSP activities under different scenarios' (Mbabu *et al.* 2015: 18). This reflects a broader tendency for OFSP programmes, like many other contemporary development efforts, to rely more heavily on the use of isolated success stories and other promotional devices than comprehensive and systematic, longitudinal field data.

Closer examination of the ways in which OFSP has been promoted in Tanzania suggests that widespread production of OFSP will depend on familiar actors and strategies (various awareness-raising and training activities, dissemination strategies, subsidies for equipment and biological material, interactions with agricultural extension staff with insufficient logistical support). We note that all of these are associated with well-documented tensions around financial sustainability, and are likely to encounter the same obstacles to long-term success as other agricultural interventions in Africa. OFSP is clearly a technology resulting from development-oriented agronomy; although contestation around OFSP is not always evident, the changing, diverse and often contradictory approaches reveal the inherent tensions within the 'biofortification for nutrition' narratives and approaches.

Notes

1   Fieldwork, which included semi-structured interviews, and participant observation, was conducted in Hai and Mwanza districts in Tanzania between June 2015 and January 2016. Interviewees included farmers, researchers from national institutes, NGO staff, and donors.

2   Initial trials were based on retinol stores in blood samples, and could not differentiate between plant based (carotenoids) and meat sources (retinol) of vitamin A.

3   Interview with NGO personnel, Arusha, 22 June 2015.

4   Tanzania Agricultural Productivity Programme took place between 2010 and 2015 and focused on strengthening value chains for potentially high market value crops including OFSP.
5   First author's unpublished data from participant observation.
6   Interviews with farmers in Ukerewe District and Misungwi District, Mwanza, October, 2015.
7   Interviews with farmers in Ukerewe District and Misungwi District, Mwanza, October, 2015.
8   Interviews with women's farmer group, NGO staff, Mwanza District, Mwanza Region, October 2015.
9   Interview, Hai District, Kilimanjaro Region, June 2015.

9

CROPS IN CONTEXT

Negotiating traditional and formal seed institutions

Ola T. Westengen

Introduction

Improved seed was the iconic technology of the Green Revolution. To some, the Green Revolution's seeds are still 'miracle seeds', while to others they are a symbol of paradise lost. Perhaps the best-known contestations involving seed are around GMOs and intellectual property rights (IPR), but the 'seed struggle' has many other battle lines. This chapter is about knowledge politics in discussions about seed systems and seed system development. The term seed system refers to the institutional arrangements involved in the supply and sourcing of seed, from variety development to seed use on the farm. It is common to distinguish between *formal* and *informal* seed systems. A formal seed system includes the chain of public and private sector activities and institutions that produce, release and distribute seed of officially registered crop varieties. An informal seed system on the other hand covers a number of different seed activities and channels, and includes saving seeds from own harvest, farmer-to-farmer seed exchange, and purchase from local markets (Almekinders *et al.* 1994; Thiele 1999; Louwaars and de Boef 2012). The scholarly literature on seed systems commonly stresses that the formal–informal distinction should be seen as a heuristic device since empirical work shows that there is no clear boundary between the two, especially in developing countries (Almekinders *et al.* 1994; Coomes *et al.* 2015). However, as this chapter will demonstrate, the formal–informal framing continues to influence debates as well as policy and practice in seed system development.

Crop improvement plays a central role in the current push to create a Green Revolution for Africa and develop 'Climate Smart Agriculture'. In addition to the technical focus, there is much policy-level work to reform laws and regulations in order to enable the development of more commercially oriented seed supply systems – e.g. by the G8's New Alliance for Food Security and Nutrition and the

Alliance for a Green Revolution for Africa (AGRA).[1] The political economy represented by this formalisation and modernisation approach to agricultural development is contested (Scoones and Thompson 2011). A common oppositional framing is promoted by social movements and NGOs advocating for local and traditional seed production and supply arrangements – e.g. La Via Campesina and the International Planning Committee for Food Sovereignty (IPC).[2] Some support for this position is also found in the academic literature (e.g. Wittman 2009; Kloppenburg 2010; Wattnem 2016).

This chapter proceeds as follows. First, it presents two case studies of seed systems, one relating to sorghum in South Sudan and the other to maize in Tanzania. These case studies are built on both genetic diversity surveys of key crops and socio-economic and cultural information on seed management and sourcing. The chapter then looks critically at the two dominant framings of seed systems in popular and scholarly debates. These framings are examined to determine whether they are reflected in the two case studies, and how they influence seed system development policy and practice. The last part of the chapter explores how theoretical approaches developed to study institutions involved in social-ecological systems can contribute to the understanding of seed systems. The argument is that critical institutional analysis can help move the field beyond the formal–informal binary to develop context- and crop-specific evidence on seed-related problems and ways that seed systems can be strengthened.

Case 1: A resilient sorghum system in Lafon

The first case describes the sorghum seed system in the Lafon villages in Eastern Equatoria state, South Sudan. The six Lafon villages are inhabited by about 30,000 people belonging to a distinct Nilotic ethnolinguistic group known as the Pari. Anthropologists characterised the Pari's economy as 'multiple subsistence', of which sorghum cultivation is an essential pillar, supplemented by livestock husbandry, hunting, fishing, and collection of wild food (Kurimoto 1984; Takei 1984). The road to Lafon is only passable by vehicle during the peak dry season and geographical isolation and recurring conflicts have contributed to the persistence of the traditional Pari way of life. As with many Nilotic ethnolinguistic groups, the age-grade system is an important social institution. Young men are enrolled in age-sets, and these pass through different age-grades over the members' life times (Kurimoto 1995). The age-grade system has critically important political, legal, military, ritual and economic functions, and also plays a critical role in the governance of the traditional sorghum seed system (Westengen *et al.* 2014a). The ruling age-grade, the *mojomiji*, decides when people should sow their fields and all Pari lineages ritually mix a bowl of seeds from a central granary with their own seeds before sowing. This practice connects sorghum fields and granaries in Lafon, creating a genetic meta-population that I refer to as a 'landrace complex'. The common name for the Pari sorghum landrace complex is *nyithin*, meaning 'the small one' and Pari informants as well as people from neighbouring groups referred to *nyithin* as the sorghum that 'came with the Pari'. In a genetic diversity study of *nyethin* we found that the landrace complex consisted of a number of genetically distinct landraces and was, as a whole, genetically differentiated from the sorghum of neighbouring groups (Westengen *et al.* 2014a). A study comparing *nyethin* collected in 1983 with *nyethin* cultivated in 2010–2013 showed no evidence of loss of diversity, despite the environmental and social stress that characterised this period. The sorghum seed system is the backbone of local livelihoods and it is a key component of the Pari society's adaptation to the local agroecology. Based on the resilience of the landrace complex and the seed system that underpins it we argued that the sorghum seed system could be considered as a 'successful social-ecological adaptation' (Westengen *et al.* 2014a).

Case 2: Creolisation of maize in Mangae

The second case relates to the maize seed system of a community living in the semi-arid drylands near Morogoro, Tanzania. The political and security situation in Tanzania is very different from South Sudan, and in terms of seed system development, the organisations, laws and regulations that typically make up a formal seed system are in place. A study of the maize seed system in the village area of Mangae was undertaken during the 2010 growing season. The local seed system was characterised by collecting livelihood data from farmers, interviewing actors from the formal system, and genotyping samples of seeds from the formal and informal maize seed systems. Categorising the seed sources as either informal or formal according to Sperling *et al.* (2008), it was found that the largest share of maize seed was sourced through informal channels, while 24 per cent was sourced through formal channels (Westengen and Brysting 2014). Much of the seed sourced through informal channels originated as modern varieties that at some point had entered the local seed system. In the formal seed system terminology such seeds are called 'farm-saved seeds'. Furthermore, households in Mangae practised seed selection: 72 per cent of the maize-growing households reported that they normally selected seeds for the next year's planting according to criteria like drought tolerance. Westengen *et al.* (2014b) presented the results of a molecular study which 'followed' the two modern open pollinated varieties (OPVs), *Staha* and *TMV1* through the local seed system. These varieties were developed by the National Maize Research Program (NMRP) based on CIMMYT material provided in the 1970s and 1980s (Moshi and Marandu 1985), and were still the most widely cultivated modern varieties in 2010. Breeders' seed (the original population developed by the breeder) was supplied by the National Agricultural Research Institute to the seed production farms of the Agricultural Seed Agency (ASA). ASA is a semi-autonomous body under the Ministry of Agriculture, Food Security and Cooperatives with the mandate to produce, process and market seeds of public varieties. Certified seed is produced on its own seed farms, as well as by small-scale farmers located in various parts of the country under a Quality Declared Seed (QDS) scheme (Ngwediagi *et al.* 2009). Some of the seeds are distributed by private seed companies, while ASA also distributes and markets seed. Quality control and certification is the responsibility of the Tanzania Official Seed Certification Institute (TOSCI). For the molecular study, seed sampled from different points in the formal seed system – from breeders' seed to commercial seed – was genotyped, as was farm-saved seed. The analysis of the genetic structure and differentiation between seed-lots indicated that the seed supplied by the formal system was indeed true to type. When the OPVs entered the local seed system through farmer seed saving, a genetic effect was visible. Seed-lots of the OPVs sampled on-farm after one or more growing seasons were significantly differentiated from those sampled from the formal system. The genetic differentiation of the farm-saved seed confirmed that hybridisation and possible local adaptation and/or selection was taking place. The conclusion was that with access to drought tolerant OPVs, farmers might be able to adapt maize to the effects of climate change (Westengen *et al.* 2014b).

Framing seed systems

How are seed systems framed in popular and scholarly debate and what kind of development approaches are supported as a result? Claude Lévi-Strauss wrote about the human propensity for thinking in 'binary opposites' (Lévi-Strauss 1955), and other structuralists have since noted how secondary binaries reinforce meanings created by the original binary, such as 'good' and 'bad'. I contend that the formal–informal seed system dichotomy is connected to such a binary; one group of actors promotes formalisation of seed systems while another group promotes maintenance of the informal system. In what follows these two framings will be referred to as modernisation and localisation, respectively.

Modernisation is arguably the dominant framing of agricultural development among national governments, funders and researchers involved in international agricultural development, while localisation is common among some scholars, activists and NGOs associated with the agroecology and food sovereignty movement. The modernisation framing is characterised by a belief in the transformative role of technology (improved seed and other agro-inputs) to increase productivity in the agricultural sector (Thompson and Scoones 2009). Framed by modernisation, formalisation of seed supply systems is necessary to facilitate farmers' access to improved seed: importantly, improved seed is portrayed as scale neutral, meaning it can benefit small farmers as well as larger farmers. Two of the development efforts that currently best exemplify this framing are AGRA and the G8's New Alliance for Food Security and Nutrition.

The localisation framing contests several of these tenets, including the scale neutrality of much modern technology and the view that increased productivity of some staple crops translates into improved food security at the household level and social and economic development at the national level (Ellis 2000). Localisation is associated with promotion of local political and economic control over food production and consumption, and reliance on local resources. In this framing, formalisation of seed systems represents a potential threat to local sovereignty. The development actors that most actively use this framing are the international farmer organisation La Via Campesina and the International Planning Committee for Food Sovereignty (IPC). In the next sections these two framings are explored in more depth, in order to address the questions: How are they reflected in the case studies, and how do they influence policy?

The current motorway: modernisation

The pathways approach to analysing contestation within agronomy is based on the recognition that unequal power relations allow some development pathways to become dominant while others remain marginal and obscure (Sumberg and Thompson 2012). The technological and institutional development connected to seed is a good example of how the modernisation framing has been used to promote a particular vision of agricultural development under the banner of the new Green Revolution for Africa (Thompson and Scoones 2009). A clear example can be seen in AGRA's assessment of seed systems in African countries (Table 9.1). This assessment placed seed systems in one of four development stages, based mainly on

TABLE 9.1 Seed system development stages defined on the basis of degree of formalisation and commercialisation by the Programme for Seed System Development of the Alliance for a Green Revolution for Africa[3]

Development Stage			
1	*2*	*3*	*4*
• No original breeding • No formal variety release process • No private seed companies • No or very few agrodealers • No outside seed investors • Limited farmer awareness of improved seed	• Some original breeding • Few small or med. seed companies • Var. release formalised • Growing agro-dealer network • Evolving seed policy environment • Early stage outside investors	• Strong breeding systems • Many small seed companies • Significant policy issues, esp. foundation seed policies, preventing further growth • Outside investors showing reluctance	• Robust breeding pipeline • Multiple stable seed companies • Strong interest from outside investors • Favourable seed policies • High farmer awareness
South Sudan Liberia Sierra Leone	Niger Mozambique Rwanda Mali	Burkina Faso Ghana Ethiopia Tanzania Nigeria	Uganda Zambia Kenya Malawi

the degree of private sector involvement. AGRA's model seed system tends toward the commercial systems found in the market economies in the global North, such as the US and EU. This framing is a direct expression of modernisation theory with its underlying premise that 'underdeveloped' regions can be developed by transferring the model of economic and social development that served the industrial economies.

Using these criteria, South Sudan is in Stage 1 at the bottom of the seed system development ladder. The sorghum case study from Lafon provides a good example of a 'pre-modern' seed system. There have been many plans and programmes for seed system development in South Sudan, but they have rarely been scaled-up beyond single interventions. The conflict that erupted in December 2013 stopped most long-term agricultural development programmes. Efforts to deliver improved seed through emergency aid have largely failed and the informal seed system remains far more important than the formal system (Jones *et al.* 2002; Westengen *et al.* 2014a; McGuire and Sperling 2016). The sustainability challenges associated with emergency and development-oriented seed system interventions, as well as the key role played by informal systems, are commonly noted (McGuire and Sperling 2016).

It is clear that the modernisation framing is a far cry from the picture that emerges from the South Sudan case study. First, the sorghum seed system in Lafon is entirely traditional and it is entangled with a cultural and social institution that has nothing to do with market economy. Second, efforts to establish a formal seed system would have limited chance of success in the current political and economic context. It is a basic insight in political science that without inclusive political and economic institutions in place at the state level, specialised sector institutions are unlikely to function (Acemoglu and Robinson 2012; Fukuyama 2014). The sorghum seed system in Lafon therefore illustrates the limitations of the modernisation approach in situations of conflict and political turmoil. Even if South Sudan is an extreme example, conflict and political instability are observed in a number of African countries, and the case therefore provides relevant lessons about the crucial importance of the default local seed systems.

The situation in Tanzania is very different. In AGRA's view, Tanzania's seed system was already at Stage 3 when the case study was undertaken in 2010. In recent years Tanzania has taken some significant steps toward developing a commercial seed industry and these developments had a real impact on the ground in Mangae. However, to date the modernisation is partial and the resulting system is of a hybrid nature. Improved maize seed was available to the farmers, but out of the 24 per cent sourced through formal channels only 9 per cent were from commercial agrodealers (Westengen and Brysting 2014). The combination of on-farm seed saving and seed selection creates possibilities for hybridisation and creation of the new genotypes combining genetic material and traits from local and modern varieties. Studies from Mexico have shown that smallholders deliberately allow such on-farm crossing – a phenomenon known as creolisation (Bellon and Risopoulos 2001). Creolisation represents one form of farmer innovation and the literature documents many other examples of farmer manipulation of genetic material. These examples demonstrate clearly the limitations of the view that only formal breeding programmes can do crop improvement (Bellon and Brush 1994; Richards 1996; Louette *et al.* 1997; Jarvis and Hodgkin 1999; Teshome *et al.* 1999; Gibson *et al.* 2005; Mekbib 2006; Richards *et al.* 2008).

A new generation of Green Revolution actors have challenged the formal system that developed and disseminated the OPVs in the Mangae area. The period between the first Green Revolution efforts and the current initiatives was characterised by limited investment in agricultural development and the dismantling of the public sector research and extension organisations (IAASTD 2009; World Bank 2007). This was part of a general trend during the era of structural adjustment in the 1990s. Starting in the 1990s the Tanzanian government took several initiatives to encourage private sector involvement in the seed sector. The legal and regulatory framework was reformed and several new laws were passed including the Plant Breeders' Rights Act (2003), which established Plant Variety Protection (PVPs) and the Seed Act (2003) which governs certification and sale. The state owned seed provider Tanseed which had collapsed in the 1990s, re-emerged in 2002 as the private company Tanseed International Ltd. Also other private companies entered the seed market around this time, including multinationals with a focus on hybrid maize. However, the private sector never filled the space left after the dismantling of public agricultural R&D and extension services (IAASTD 2009), nevertheless, the emergence of the new Green Revolution agenda has re-energised private actors to advocate for the formalisation of the seed system.

The contrast between the early and more recent interest in Green Revolution is well illustrated by the seed company Tanseed. As in the case of the public OPVs used in Mangae, the breeding lines that Tanseed uses to develop maize varieties are from CIMMYT. But unlike the public OPVs, the new Green Revolution varieties are protected by PBRs (Plant Breeders' Rights) and distributed through a fully commercial pipeline. In order to provide conducive conditions for development of the seed sector the government has implemented some important legal and policy measures. Tanzania was the first Least Developed Country (LDC) in the world to join the 1991 International Convention for the Protection of New Varieties of Plants (UPOV91), a strong legal regime for Plant Variety Protection and Plant Breeders Rights. The 2003 Seed Act (amended in 2014) is also set to be replaced with a new UPOV91 compliant law. It is notable that a recent article in the *Journal of Intellectual Property Rights* concludes that the PVP protocol adopted in Tanzania goes 'beyond UPOV91 requirements' and 'gives stronger rights to the breeders than those found in the most advanced industrialised countries' (Haugen 2015). It thus appears as if, at least on paper, seed sector modernisation in Tanzania has surpassed the model in terms of providing for private sector interests. In practice, however, the formalisation agenda seems to have had limited traction so far. Mkindi (2015) cites an AGRA report saying that the formal system only makes available between 10,000 and 15,000 tons toward the 120,000 tons annual seed requirement.

The counter-movement: localisation

The most vocal criticism of the modernisation framing comes from activists and scholars associated with the agroecology and food sovereignty movements. Agroecology has become the 'technological flag of the resistance movement' (Wittman 2009) The IPC in 2015 published the 'Declaration of the International Forum for Agroecology'.[4] Like AGRA's development stages, this declaration illustrates a strong normative framing of seed system development. The declaration says: 'Farmers' seeds are being stolen and sold back to us at exorbitant prices, bred as varieties that depend on costly, contaminating agrochemicals' (p.2). The IPC declaration goes quite far in saying that there is no room for scientifically bred seeds in the kind of seed systems they promote. In place of the modernisation pathway they argue for a counterstrategy to 'Take back control of seeds and reproductive material and implement producers' rights to use, sell and exchange their own seeds and animal breeds' (p.5). According to the declaration, local seeds under the control of community-based seed systems represent the desirable alternative to formal seed systems distributing proprietary varieties. La Via Campesina has promoted this notion in campaigns for 'seed sovereignty' (Wittman 2009). Wittman describes the rationale behind a campaign in Brazil this way:

> Farmers in Mato Grosso have found that regional varieties of seeds (especially beans) have much higher yield and require less pest management than hybrid or non-local seeds promoted by local seed companies and government extension agents. This recognition of local ecological limits to seed viability and production has informed the MST [Movimiento de Trabajadores Rurales Sin Tierra] seed campaign, which encourages local 'seed sovereignty' or control over local seed and genetic resources as a direct response to the increasingly limited seed sources controlled by agribusiness.
>
> *(Wittman 2009: 817)*

The organisation African Forum for Food Sovereignty (AFSA) and the NGO GRAIN are explicit in their critique of the current efforts to formalise African seed systems:

> Privatising both land and seeds is essential for the corporate model to flourish in Africa. ... With regard to seeds, it means having governments require that seeds be registered in an official catalogue in order to be traded. It also means introducing intellectual property rights over plant varieties and criminalising farmers who ignore them. In all cases, the goal is to turn what has long been a commons into something that corporates can control and profit from.
>
> *(GRAIN 2015: 2)*

Returning to the question of how the case studies reflect the framings, and taking seed sovereignty to mean 'control over local seed and genetic resources' (Wittman

2009), one could argue that the Pari community, with no form of official regulatory regime, is a rare example of a seed sovereign community. The more relevant question though, is whether seed sovereignty is a concept that resonates with the Pari themselves. This question requires a different kind of study, but it is questionable if it makes sense to speak of seed sovereignty for a community isolated by warfare. To what extent the seed system in Mangae reflects the localisation framing and its normative agenda is a question of interpretation. The OPVs circulating in the local seed system were developed from a broad germplasm base at CIMMYT and the varieties are indeed registered in the national seed catalogue. Thus if seed sovereignty implies reliance on local seed (i.e. traditional local landraces) and absence of official seed regulations, the community in Mangae cannot be said to be sovereign. However, a position based on such a narrow definition of seed sovereignty is probably mainly reflected in theoretical discussions and it seems the counter-movement actors actively involved in struggles about the future of Tanzanian's seed system have a broader understanding of seed sovereignty. A recent report from the African Center for Biodiversity (ACB 2015) on Green Revolution interventions in Tanzania used strong rhetoric against the private sector interests and was critical of AGRA's approach to seed sector development. Nevertheless, the report recognises that crop improvement and public sector seed systems can have positive development dividends: 'AGRA's interventions raise a number of issues for the food sovereignty movement, in particular they impel us to clarify our positions on public and private sector R&D, germplasm improvement, and the role of farmers in seed production and distribution' (ACB 2015: 21). On the issue of what kind of seeds should be encouraged the report includes improved OPVs in addition to local varieties:

> The first question is whether improvements in genetic materials are required. There may be an argument that local germplasm and varieties are well suited to local ecological conditions and do not need to be supplemented with materials from outside Tanzania's borders. However, external genetics are usually brought in to improve selected traits in local materials, including yield enhancement, pest and disease resistance, drought or salinity tolerance, etc. These genetic improvements may be beneficial and evidence of this is farmer adoption.
>
> *(ACB 2015: 46)*

The ACB position is therefore one that rejects a model that necessitates exclusive rights to or ownership of genetic material and maintains that only the public sector should be involved with seed sector development.

There is undoubtedly less money and power behind the localisation compared to the modernisation movement. Nevertheless, the oppositional framing is influencing the development of seed systems in Tanzania in both subtle and direct ways. NGOs and social movements are opposing the Green Revolution model in the media and by promoting alternative approaches in the field. At the regional level the Alliance for Food Sovereignty in Africa and other civil society organisations have protested to the African Regional Intellectual Property Organization (ARIPO) about the process and the adoption of the new PVP protocol (AFSA 2014; Haugen 2015). As a response to the protests the regional policy protocol included the text 'the situation of smallholder farmers will be taken into consideration in relation to farm-saved seeds' (ARIPO 2013: 210, cited in Haugen 2015). In Tanzania, members of the alliance against the regional protocol such as the Tanzania Alliance for Biodiversity (TABIO) have come out strongly against the country's new PBR regime and in favour of seed sovereignty:

> By creating laws to provide incentives to companies to develop high-yielding varieties of seeds, these laws hand over the control of seeds and by proxy the country's seed system to seed corporations that are then free to exploit farmers by gradually replacing traditional seed with a uniform and limited number of commercial products that cannot be saved or traded. This control of the seed system by a handful of companies will be disastrous for small-scale farmers and the biological diversity that traditional seed varieties uphold in Tanzania.
>
> *(Mkindi 2015: 3)*

Binary misconceptions?

As outlined in the previous sections, Tanzania is now about to establish policy and legal regimes of the kind that proponents of modernisation say is necessary to enable a well-functioning seed sector, and opponents say will harm the seed systems smallholders rely on for their food security. It is not within the scope of this chapter to assess the implications of these changes, but the debates surrounding these recent developments cast light on the knowledge politics in seed system development. The different ways of seeing from each side of the formal–informal binary are also reflected in perspectives on PVPs and seed laws (Table 9.2).

TABLE 9.2 Binary opposites in debates over plant variety protection and seed regulations

Modernisation	Localisation
• Intellectual Property Rights in the form of PVP and PBR are necessary for innovation to happen.	• Seed laws and PVPs are basically the same and their purpose is to protect the ownership of genetic resources for the private seed industry.
• Trust in seed transactions relies on certification.	• PVP and compliant seed laws denies seed saving, exchange and sale of all varieties.
• Strict PVP and a new Seed laws will enable the private sector to move in and supply seeds to a ready demand from farmers.	• The seed struggle is basically the same in all parts of the world.

The modernisation framing is arguably based on a turn toward standardisation for which there is limited empirical support from developing countries. There is reason to problematise all three assumptions and premises on the left-hand side in Table 9.2. First, it is not true that innovation requires IPRs. Both technological and institutional innovation take place in farmers' fields and in public research institutions in the absence of PVPs or PBRs. The OPVs dominating in the case study area in Tanzania were publicly developed based on both international and local genetic material, and public ownership did not hinder either commercial actors multiplying them for sale or farmers saving, exchanging and selling them. Second, local markets are a major source of seed and farmers' use of local markets for seed indicates a certain level of trust (Sperling and McGuire 2010). If there is mistrust, it is likely related to sale of counterfeit certified seeds. The third assumption about what hinders farmers' access is also not supported by empirical evidence. Haug *et al.* (2016) found that good quality improved seeds exist and are available through agrodealers, and that the problem of access is mainly about the 'enabling environment' rather than the supply system itself. Related to this misconception is the 'widespread perception' among the private sector and some donors that competition from the public sector constrains private sector seed sales (Haug *et al.* 2016). It is to state the obvious, but the main reason why farmers cannot access seeds is poverty. The modernisation view on PVPs and seed laws is an example of 'seeing like a state' (Scott 1998) while the perspective of the smallholder is likely to be quite different. In a study of the political ecology of farmers' seed decisions in three West African countries, Jones (2013: 16–17) states: 'For many farmers in West Africa, the social and natural contexts shape a different set of priorities, needs and opportunities for the use of improved variety seeds.'

As Lévi-Strauss observed, a common reaction to a framing we disagree with is to turn to its binary opposite (Lévi-Strauss 1955). The real stories are usually more complex and messy than this, yet binaries continue to play a major role in narratives supporting localisation. The assumptions on the right side of Table 9.2 are examples of tropes from such narratives, and it is important to consider them critically. The classical locus of seed struggles has been intellectual property rights regimes with their patents and PBRs, and the political economy scholarship on seeds has until recently paid scant attention to national seed laws (Wattnem 2016). However, while briefly recognising the role of seed laws in ensuring seed quality, scholars in this tradition also extend their criticism to this part of the formal system, arguing that seed laws and regulations provide another set of mechanisms for 'accumulation by dispossession' (Kloppenburg 2010; Wattnem 2016).

It is important to distinguish clearly between PVPs on one side and seed laws and regulations on the other. The origins of seed laws lay in the need to ensure seed quality in formal seed system transactions, i.e. to protect farmers from seed that are not true to type, of low viability, or are contaminated by weeds etc. It is widely acknowledged that poor quality is a problem for farmers purchasing seeds from formal seed systems in sub-Saharan Africa, and seed laws are arguably a necessary element in a functional formal system (Langyintuo *et al.* 2010).

PVPs, on the other hand, are about ensuring IPRs, based on the notion that owners of a new plant variety need an exclusive right to capitalise on it in order to encourage investment in plant breeding. Thus, PVPs are about protecting ownership and they are much more important in a commercial formal system than they are for public formal institutions (Tripp 1997). But also with regard to PVPs it is important to be accurate and the second point reflects a common misunderstanding about the purview of such laws. PVPs like UPOV91 will protect only new varieties that are granted PBR. Thus varieties that are not registered for PVP, such as landraces and modern varieties released earlier should remain free of restrictions on their use, exchange and trade. Many localisation narratives skim over the fact that PVPs only apply to protected varieties: this contributes to the polarisation of the debate, i.e. that it is about either only formal or only informal systems. The third and last point on the right side in Table 9.2 is the response by those promoting the localisation framings to the 'one size fits all' thinking that too often accompanies the modernisation framing. There are heated debates in Europe and other developed countries over seed regulations, but the arguments used cannot easily be transferred to debates in countries in the global South. There is for example a fundamental difference between the situation with farmers in Europe who fought to register heirloom varieties in national and regional seed catalogues and the situation in developing countries where more than 80 per cent of the seed is sourced from the informal system and where most are local varieties. The trade in heirloom varieties in a European country is a small niche within a highly formalised seed system, and the process of amending the EU seed regulations to make it possible to have heirlooms varieties registered is not necessarily a good model for developing countries. Registering a variety is arguably a form of privatisation and it is unclear what benefits would arise from this kind of formalisation within a developing country seed system.

Seed systems as social-ecological systems

If the formal and informal framings are lacking, what are the alternatives? Perspectives from social-ecological system research (SES) (Berkes *et al.* 2000; Redman *et al.* 2004) and social and ecological resilience frameworks (Adger 2000; Folke 2006) represent interesting vantage points for seed systems research. These focus on social-ecological dynamics on different but interlinked spatial, temporal, and organisational scales, and in relation to different drivers of change.

The two case studies described earlier can be used to illustrate how seed systems can be conceptualised as social-ecological systems. In the case from Lafon the term 'social-ecological adaptation' was used to describe how social and ecological factors are woven together in the Pari seed system. The notion that cultural practices are adaptations to the local environment has its antecedents in cultural ecology (Smit and Wandel 2006) and is today represented in various scholarly traditions on human-environment dynamics (Turner and Robbins

2008). Seeds are manifestations of environmental and cultural adaptations and seed systems are clearly social-ecological systems that encompass relationships between people, crops and the local agroecology. The maize seed system in Mangae exemplifies a common pattern in smallholder economies in developing countries where farmers have one foot in the traditional economy and the other in the market economy. This is no less a social-ecological system than that in Lafon, but the market economy exercises another set of political and economic influences on the system.

There are several SES research approaches that can be useful in the study of seed systems, but here the focus is on a tradition that resonates particularly well with the empirical evidence from the case studies. This tradition can broadly be categorised as 'institutional analysis' (Nunan 2015), with institutions understood as 'the rules of the game' and encompassing national policies and legislation as well as local rules and norms. Elinor Ostrom's Institutional Analysis and Development (IAD) framework has been widely used to analyse how institutions and collective action have evolved to govern common property (Ostrom *et al.* 1994; Ostrom 2015). This framework could be useful for analysing the tension between traditional approaches to the management of seed and the IPR and transaction regulations introduced with formal seed system development. However, the creolisation in Mangae challenges the application of universal models for the analysis of collective action. As a response to the complex and messy nature of property rights in developing countries, the original IAD/SES frameworks have evolved considerably and 'critical institutionalism' has developed as a less prescriptive and more explorative approach that 'gives greater recognition to institutions that mediate access to, and control over, natural resources but are not necessarily "designed" or "developed" with natural resource governance in mind' (Nunan 2015: 63). This type of institutional analysis is thought to be better able to grasp 'development through bricolage': 'a process in which people consciously and non-consciously draw on existing social formulae … to patch or piece together institutions in response to changing situations' (Cleaver 2012: 45).

The concept of bricolage used in this connection was originally developed by Lévi-Strauss. French agriculture used to have so-called bricoleurs – handy-men or 'Jacks-of-all-trades' – who tinkered and created solutions with whatever materials were available. Lévi-Strauss used the concept of bricolage as a metaphor to describe how people think in pre-modern societies:

> The 'bricoleur' is adept at performing a large number of diverse tasks; but, unlike the engineer, he does not subordinate each of them to the availability of raw materials and tools conceived and procured for the purpose of the project. His universe of instruments is closed and the rules of his game are always to make do with 'whatever is at hand', that is to say with a set of tools and materials which is always finite and is also heterogeneous because what it contains bears no relation to the current project, or indeed to any particular project, but is the contingent result of all the occasions there have been to

renew or enrich the stock or to maintain it with the remains of previous constructions or destructions.

(Lévi-Strauss 1966: 11)

In relation to seeds, the bricoleur is the farmer who does her/his best with the genetic resources at hand, while the engineer is the professional plant breeder. This is not a bad analogy for the processes we see at work in the case of the maize seed system in Mangae. Furthermore, Cleaver's (2002, 2012) distinction between bureaucratic and socially embedded institutions is arguably more suitable than the formal–informal dichotomy: 'Bureaucratic institutions are those formalised arrangements based on explicit organizational structure, contracts and legal rights, often introduced by governments or development agencies. Socially embedded institutions are those based on culture, social organization and daily practice' (Nunan 2015: 14). The institution involved in governing the sorghum system in Lafon is a good example of a socially embedded institution. The age-grade system is fundamentally important to social organisation in the Pari culture and the fact that seed is also governed through this system actually makes seed management very formal. The connection between seed and the age-grade system also resonates with another important insight from critical institutionalism, namely that influence and control over natural resources often is governed by social institutions that were not necessarily designed for the purpose (Cleaver 2012). Also in the case of creolisation in the maize seed system in Mangae the critical institutional perspective provides for some interesting insights. While farm-saved seed is commonly understood as a consequence of poor farmers' inability to purchase new seeds every year, it can also be seen as an example of development through bricolage (Cleaver 2012) and *metis* (Scott 1998). The farmers draw on technological options from both bureaucratic and socially embedded institutions and sometimes expand on these options by combining them in new ways. This is an example of the adaptive nature of development through bricolage – institutions evolve. In this case, the formal seed development and supply system delivers the technology it is supposed to deliver, but the reach of the formal system ends there. The seed laws that are supposed to regulate use and trade are not enforced and the bureaucratic institutions are less rigid in practice than in theory.

Studying seed systems through the lens of critical institutionalism allows us to see both social and environmental sustainability outcomes. This perspective can contribute to the contemporary turn toward a more smallholder-centred seed and food security focus in the seed system literature. While much of the early literature on seed systems focused on the ability of local seed systems to conserve genetic diversity (*in situ* and on-farm) recent contributions more often focus on their role in supporting adaptation and resilience in the face of external stress (Bellon and Hellin 2011; Bellon *et al.* 2011; McGuire and Sperling 2013; Westengen and Brysting 2014). Moreover, critical institutionalism offers a lens that is sufficiently wide to see that modernisation and localisation not are mutually exclusive, nor the only ways of seeing. The argument is that critical institutionalism has the potential to open up space for alternative narratives and plural pathways as called for in the introduction chapter and in earlier contributions drawing on the pathways approach (Thompson and Scoones 2009; Scoones and Thompson 2011; Sumberg and Thompson 2012).

There are many alternative ideas about and approaches to seed systems. Some are well established and others are still on the drawing board. Participatory plant breeding is a well-established alternative approach to crop improvement, yet it remains relatively marginal. And while private seed companies continue to expand their market share globally the tendency is that conventional plant breeding in the public sector continues to be sidelined. The Open Source Seeds Initiative is an attempt to create a game-changing counter-movement to the dominant and expanding IPR regime (Kloppenburg 2014). Other authors advocate for a pragmatic approach, promoting coexistence of the commercial seed sector with non-commercial seed systems relying on PVP rules that allow smallholders to use, save and exchange seeds (De Jonge *et al.* 2015). The Quality Declared Seeds scheme is an established alternative to centralised seed production in many countries and Ethiopia is currently operationalising an integrated seed system approach (Louwaars and de Boef 2012). Furthermore, NGOs have for many years supported various alternative approaches to variety dissemination such as 'seed vouchers and fairs' (Remington *et al.* 2002) and community seed banks (Vernooy *et al.* 2015). These and many other approaches pursued by farmers themselves can lead to both technological and institutional innovations and represent alternatives to the dominant pathways for seed system development. In the face of the current challenges to seed and food security there can be little doubt about the fact that different approaches are needed to different problems, for different crops, and in different places.

Notes

1 The New Alliance for Food Security and Nutrition has as one of its Commitments: 'Policies that regulate the production, distribution and use of improved seed, fertiliser, pesticides and farming implements' (https://new-alliance.org/commitments). AGRA is involved in seed system development through its Programme for Africa's Seed Systems (PASS) (http://archive.agra.org/what-we-do/seed/pass-subprograms/)

2 La Via Campesina is part of the IPC and their position on seed system development is e.g. expressed in the Declaration of the International Forum for Agroecology (http://www.foodsovereignty.org/forum-agroecology-nyeleni-2015/)

3 Table adopted from AGRA PASS presentation, available at: http://www.fao.org/fileadmin/user_upload/drought/docs/AGRA%20Seed%20Systems%20and%20the%20future%20of%20farming.pdf

4 http://www.foodsovereignty.org/forum-agroecology-nyeleni-2015/

10

LAWS OF THE FIELD

Rights and justice in development-oriented agronomy

James A. Fraser

Introduction

Development-oriented agronomy aims to address so-called 'global grand challenges', such as improving food security and wellbeing, reducing poverty and climate change adaptation and mitigation (e.g. Alliance for a Green Revolution in Africa). Indeed, food and agriculture lie 'at the heart of' the 17 Sustainable Development Goals (as emphasised at the FAO's 43rd Committee on Food Security in October 2016). This convergence between the fields of development and agronomy opens up new forms of contestation right across the agendas, framings, narratives, partnerships, methods, results, and impact evaluation of agricultural development in 'the South' (Andersson and Sumberg 2017). This has been shaped by three important changes in agricultural research since the mid-1970s: the increasing influence of neoliberalism, environmentalism, and participation on agricultural development processes in the South (Sumberg *et al.* 2013). These changes have influenced critiques of agronomy around for example the merits of local versus scientific knowledge (Richards 1985); conventional versus agroecological systems (Kuyper and Struik 2014); and small versus large farms (Collier and Dercon 2014). This chapter proposes that one underlying reason for this contestation is that as agronomy – a natural science – has become more oriented to development, it has been opened up to *normative* contestation within and between multiple arenas, including developing world contexts of technology transfer where negative effects on human rights are challenged by NGOs and civil society; through ontological and epistemological frictions with social scientists; and in the new policy arenas that agronomists navigate.

Giving the rights and justice issues around development-oriented agronomy serious consideration means investigating the effects of agronomic innovations on human freedom, since rights, justice and development are, at their most fundamental level all about human freedom (Sen 2001). The contestation of development-oriented agronomy on normative grounds is not new; it began with critiques of colonial plantation systems as I discuss below, but to date this has not been adequately addressed. I make the case therefore that the particular politics of knowledge advanced by the contested agronomy project would benefit from engaging more explicitly with rights and justice. Moreover, it is important to be explicit about the divergent ways in which different underlying philosophical traditions represent rights and justice, which in turn may shape both contestation and development outcomes. This comes at a moment in the social sciences when there is a revitalisation of interest in normative theories of society and a concomitant shift from idealised models of justice to taking everyday understandings of injustice, denigration and harm as the point of departure. These locate ethics and morality in practical reason or *phronesis* as well as in detached, rational deliberation. Hence, normative concepts of justice are taken to inhere in arenas of contestation and so social justice can start from readily available societal, in addition to legal or philosophical understandings of injustice (Sen 2009; Sayer 2011, 2015; Barnett 2011, 2012, 2014; Caillé and Vandenberghe 2016).

Normative contestation in development-oriented agronomy

I contend that the contestation of development-oriented agronomy, particularly by those promoting the environment and participation agendas identified by Sumberg *et al.* (2013), has implicitly been driven, at least in part, by normative concerns around the negative social and environmental impacts highlighted in some critiques of the Green Revolution (Drèze and Sen 1989; Negin *et al.* 2009; Patel 2009; Pretty *et al.* 2009; Kerr 2012; Patel 2013; Akram-Lodhi 2013). Rights and justice tend not to be explicitly theorised, however – neither in conceptual frameworks used in development studies nor across the whole spectrum of social sciences. Rather, normative concerns are often left implicit. Given the urgency of global agricultural and broader related 'nexus' challenges, as the sociologist Andrew Sayer recently put it, 'it seems irresponsible to continue the academic tradition of avoiding normative judgements of what is good or bad, life-enhancing or life-threatening, just or unjust' (Sayer 2015: 291). Human rights principles and standards are now strongly reflected in the UN 2030 Sustainable Development Goals and the World Social Science Report 2016 focuses on justice and inequality, both of which open up the policy arena to more explicit treatment of normative concerns (ISSC, IDS and UNESCO 2016).

A corollary of making normative concerns more explicit is that questions will arise concerning the conceptualisation and measurement of 'success' in development-oriented agronomy, since simplistic metrics can mask injustice and rights violations. In a recent example from Rwanda, an 'African Green Revolution' project raised yields but only a wealthy minority benefited – for poorer rural inhabitants, landlessness, inequality and poverty were actually exacerbated (Dawson *et al.* 2016). This suggests lessons have not been learnt from the Asian Green

Revolution, where a singular focus on yield, and failure to consider and address the likely systemic effects, led, in some cases, to farm landholding consolidation and the loss of land and/or abandonment of agriculture by the poor farmers (Shiva 1991; Otsuka *et al.* 1992; Freebairn 1995; Negin *et al.* 2009; Scott 2008; Pingali 2012). Rather than just measures of yield and aggregated income, the success of agricultural development programmes must be assessed against a much broader array of 'ecological, economic and social objectives, such as sustained reductions in chronic malnutrition, poverty and ecological harm' (Thompson and Scoones 2009: 387; also see Loos *et al.* 2014). I extend this argument, by suggesting that the success or failure of development-oriented agronomy must also be assessed in normative (i.e. are its effects 'just', 'fair', 'virtuous,' 'right,' or 'good'?), as well as 'factual' (i.e. measurable, positivistic) terms (e.g. whether an intervention affected yields, calorie availability, land tenure, farm consolidation and smallholder exodus?). This also means moving beyond the dichotomy between facts (and measurement) on the one hand, and values on the other. While this distinction has long been critiqued in social science, even some in the natural sciences are now more open to such an approach (e.g. Baumgaertner and Holthuijzen 2016).

The increasing interdisciplinarity needed for this inevitably causes frictions between researchers employing positivist and interpretivist lenses. I argue that much contestation is caused by misunderstandings between those working with these different lenses. For example, a particular point of contention is over trade-offs between the successful implementation of agricultural technologies on the one hand, and the reshuffling of social structures (and the normative implications of this for rights and justice) which may be necessary to make the technologies work. These trade-offs were seen between local people's priorities and alternate wet and dry rice cultivation methods (Howell *et al.* 2015). In this particular case, an innovation which maintained rice yields while cutting the use of irrigation water by half was not taken up by Nepali farmers because it did not fit with the overall exigencies of their local water management system. Yet plant scientists continue to develop new technologies without consideration of how they fit with the social contexts into which they are to be introduced, justifying their work with self-legitimating Malthusian rhetoric about 'perfect storms' and the challenge of feeding 9 billion by 2050 (Tomlinson 2013).

Global change research initiatives, such as Future Earth[1] are doing an important job highlighting the need to co-design and co-produce research with diverse societal stakeholders, or 'users' (Mauser *et al.* 2013). There are issues however, with the ways in which Future Earth and institutions such as the Stockholm Resilience Centre and Resilience Alliance address such challenges of interdisciplinarity using the concepts of socio-ecological systems and resilience to conceptualise coupled societal and environmental changes, including agricultural systems. The issue is that this framing can lead to ontological monism – the idea that there is only one world, which can be understood by way of a single epistemological template. Similar to debates on geo-engineering as a solution to climate change, debates around plant science as a solution to food security are dangerous because of the absence of any substantive sense of ethics, justice, morality, virtue or notions of 'the good'. More provocatively, what would problems of, and solutions to, challenges in development-oriented agronomy look like 'when we examine them through complex virtues like wisdom, humility, integrity, faith, hope, and love?' (Castree 2015: 249; see also Castree et al. 2014 and Hulme 2014). Moreover, such interdisciplinary concepts, and perhaps resilience in particular, are arguably not conducive to the realisation of normative concerns of rights and justice (Bettini et al. 2016; Cavanagh 2016).

It is increasingly accepted, however, by social and natural scientists alike, that the global grand challenges (including those to which development-oriented agronomy addresses itself) are invariably 'wicked problems', to which there is no single solution. Moreover, interventions bring socially differentiating effects – benefits to some and burdens to others – and so can be both good and bad, rather than being right or wrong, correct or incorrect (cf. Rittel and Webber 1973). For example, while the Green Revolution may have doubled yields (an agronomy win) in some cases it also exacerbated inequality (a development fail). Accepting this makes serious consideration of not only social, but also ethical and moral concerns inescapable. This does encounter resistance from the so-called 'engineering mind-set' (Armstrong 2013; Ramalingam 2013) which characterises some agronomists and crop scientists. Indeed, the engineering mind-set is evident in one of the most prominent current cases of contestation in development-agronomy.

Prosavana in Northern Mozambique was initially intended to promote a large-scale 'Brazilian model' of commercial agriculture along the Nacala corridor (Clements and Fernandes 2013). Prosavana has been contested from the beginning precisely because of normative concerns emanating from Mozambican civil society and local communities. In June 2014 a national 'no to Prosavana' campaign[2] was spearheaded by Mozambican civil society and the National Peasant Union (UNAC). Following this, Prosavana's Second Masterplan of March 2015 (MASA 2015) declared it would 'improve the conditions of life' of small-scale, family producers, promising participation, respect for local culture and 'sovereignty', while supporting marginal groups like youth and women. Despite its nominally participatory and pro-smallholder approach, this new Masterplan echoed the top-down agricultural policy of the colonial era with a demand to 'break the paradigm' of smallholder shifting agriculture. This was represented in Malthusian terms – Mozambique is at a 'critical point', where population growth is outstripping land availability; change is needed to avert 'large-scale environmental destruction'. This directly contradicts the argument initially used to justify Prosavana, that there are millions of hectares of un- or under-used land along the Nacala corridor. Smallholders themselves are depicted as not being conscious of their predicament: the plan is in no doubt that they must 'change their mentality' and become 'modern', 'efficient', 'competitive' and 'market-oriented' in order to survive (MASA 2015: 28–29). This echoes a recent anecdote from a Gates Foundation project by the African Enterprise Challenge Fund that reported that a central problem they faced was that most farmers there 'viewed agriculture as a way of life and not a business'.[3] The contradictions inherent in the Prosavana case, on the one hand claiming to want to 'improve the conditions of life' while at the same time insisting smallholders undergo cultural change so that technological implementation can be successful, are part of the reason why Prosavana has been labelled 'unjust' by a range of influential actors in Mozambique, including local and regional civil society groups, farmers and the Catholic Church.[4] Currently (September 2016) a third Prosavana Masterplan is being written by the Mozambican Government with significant input from civil society including Nampula Civil Society Platform (PPSOC). The Mozambican Government apparently now accepts some of the problems it caused with previous incarnations of Prosavana.

Hence, spaces are opening up where agronomy and agricultural development more broadly are being forced to take on normative concerns. This means that the emergent field of development-oriented agronomy needs the conceptual tools to make sense of such interfaces. These kinds of national contestations over underlying ethical issues suggest that rights and justice are playing a greater role in development-oriented agronomy. But the normative contestation of agronomy has deep historical roots, however. This is related to the fact that the development of agronomy has paralleled a shift from land and resources as commons to land and resources accumulated as capital, by way of processes of enclosure beginning in the fifteenth century in the UK and continuing around the world (Capra and Mattei 2015; Moore 2015). The associated shifts from customary to formal land tenure laws have had major impacts on rights and justice in terms of the 'accumulation by dispossession' which this entailed. The origins of both development-oriented agronomy and its effects on rights and justice can then be traced through the emergence of modernity itself. On the one hand, agronomic science was made possible by the general emergence of science during the sixteenth to eighteenth centuries, with the Haber process being a critical moment (Akram-Lodhi 2013). On the other hand, in part through the transfer of agronomic science to the developing world, the origins of the current globalised capitalist food system can be traced through the colonisation of the tropics, which in many respects made possible the industrial revolution in the North (Beckert 2014; Mintz 1986).

This was followed by what David Nally calls 'an astonishing act of historical amnesia' – the rediscovery of what is now called 'the South' as lands of low productivity and hunger in need of the development's modernising project (Nally 2011a, 2011b). The amnesia arises because of the role of European coloniser states in creating the very situation that they now seek to remedy (Davis 2001). It is important to acknowledge that small famers have struggled in the face of all this, ranging from outright rebellion to everyday resistance (Scott 2008). This is not restricted to political actions but also encompasses resistance in the form of agricultural practice itself, captured in the new concept of the 'anti-commodity' where local people re-shape agro-ecologies to resist capital accumulation. An example is the way that Sierra Leonean farmers 're-engineered' Asian white 'Carolina' rice by crossing it with local red African rice varieties. This made it more adapted to local conditions but reduced its market value; both of these traits benefited local farmers but not capital (Maat and Hazareesingh 2015, see also Carney and Rosomoff 2009).

The etymology of the word *agronomy* would also appear to support a link with rights and justice. The word agronomy derives from Ancient Greek ἀγρός *agrós* 'field' and νόμος *nómos* 'law'. Law here might be taken to refer to 'the laws of nature', in the still dominant Cartesian sense. But various other traditions, from the holistic thinking of the ancient Greeks (Capra and Mattei 2015) to phenomenology (e.g. Hussurl, Heidegger), through to the current critical social sciences of the Anthropocene (e.g. Lövbrand *et al.* 2015) have sought to work beyond the nature–culture divide (albeit in different ways). Following this I take 'laws of the field' to mean both social and natural laws at play in the field of agricultural development in the South.

Bringing rights and justice into development-oriented agronomy: a framework

Rights and justice are rooted in ethics which exist in different forms in all cultures (Dussel 2013). Nevertheless, the modern application of these concepts in countries where development-oriented agronomy is carried out is influenced by Western concepts of rights and justice, which are fundamentally about freedom. But what is meant by freedom is by no means clear, and has been a central problem of philosophical inquiry since the Enlightenment (Pippin 1991). This is where how we conceive of freedom in development, and therefore what it means to different cultures around the world, becomes important. This can be approached in fundamentally different ways, via analytic, continental or post/de-colonial philosophies.

In this section I present a framework which encompasses: *Rights*, commonly understood as moral and/or legal entitlements that individual persons and groups should have, although varying from universal to particularistic definitions; and *Justice*, usually taken to mean fair, equitable and respectful treatment of people, processes and outcomes. In addition to being the subject of major disagreement among philosophers, what constitutes rights and justice varies both between different societies, and among the 'differentiated citizens' within them (Holston 2008). Rights and justice can be seen to inhere in contextual values of everyday 'ethical life' along with their codification in manifold forms of customary law across different cultures. I approach rights and justice theoretically drawing on capabilities, recognition and cognitive justice theory (Table 10.1). I have selected these three conceptual framings because they represent analytic, continental/critical and post- and de-colonial philosophies, respectively. The advantage of drawing these approaches together is that they represent Western and non-Western thought.

The Analytic tradition has dominated anglophone philosophy departments since the beginning of the twentieth century. It sees itself as complementary to, and taking the lead from, the natural sciences. Analytic philosophers take a universalist

TABLE 10.1 Three philosophical traditions

	Tradition		
	Capabilities	*Recognition*	*Cognitive justice*
Philosophical underpinnings	Analytic	Continental philosophy; Critical theory	Post-colonial; De-colonial
Focus and goal	Individual flourishing	Intersubjective respect	Intercultural dialogue in an ecology of knowledges
Link to development-oriented agronomy	As a theory of agency to look at whether interventions help individuals realise capabilities or impinge upon them	Recognition of local knowledge; participatory agenda	Engagement with social and political movements e.g. Via Campesina

approach to rights and justice in seeking abstract, trans-contextual justification of pure principles, then applying these to real contexts to assess injustice. This approach has tended to emphasise universal instrumental rationality and individual subjectivity.

Continental philosophy is more focused on local contexts and in doing so counters the universalising tendencies of forms of analytic philosophy. Continental philosophers reject the idea that the natural sciences are the only way to understand natural phenomena. While analytic philosophy sees problems as discrete from their historical origins, continental philosophy argues that problems cannot be considered apart from their context. Continental philosophers therefore tend to emphasise how reason is made manifest through socio-historical contexts and that subjectivity is a collective phenomenon.

Both analytic and continental philosophies have been critiqued by post- and de-colonial philosophers for being Eurocentric and a form of epistemic violence when applied uncritically in the South. This is because they risk imposing modes of knowing and being (expressed as modernity/rationality) (Quijano 2007) that derive from the European Enlightenment. This is particularly important given that countries targeted by development-oriented agronomy are affected by the 'coloniality of knowledge', that is, the valuing of European modes of being, knowing and doing and the devaluing of subaltern knowledges(s) (Quijano 2007). Such critiques therefore resonate with the studies of 'local knowledge' which form part of the intellectual heritage of the contested agronomy project. Engaging with post and decolonial philosophies opens up a much wider landscape of epistemological and political possibilities – moving beyond the hegemony of northern theory, analytic or continental, by way of the mobilisation of 'theory from the south'

(Comaroff and Comaroff 2012; Rosa 2014) – e.g. South American philosophies such as *Sumak Kawsay* and African philosophies such as *Ubuntu* (de Sousa Santos 2015; Gudynas 2016) – toward liberational, emancipatory ends (Dussel 2013).

Capabilities

Social scientists working on rights and justice issues in relation to environment and development are increasingly adopting the analytical capabilities approach of Sen and Nussbaum (Sen 2009; Nussbaum 2011). This approach to human wellbeing, which revolves around the different capabilities necessary for individual flourishing, has been hugely important in moving beyond simple aggregate economic indicators of development.

Capabilities emerged from entitlements theory which was developed by Amartya Sen, and which enabled him to demonstrate that simply increasing overall food production does not guarantee that fewer people will go hungry; nor does increased production equate to increased food security, which is more to do with distribution of and access to food than production *per se* (Sen 1981; Drèze and Sen 1989). Its theoretical origins are to be found in the Aristotlian notion of *eudaimonia* or 'human flourishing'. It also draws particularly on the work of John Rawls, and can be seen as within the Rawlsian tradition of liberal justice theory.

Capabilities theory turns on two linked concepts, functions and capabilities. A function is understood by Sen as the achievement of any 'state of being'. This ranges from the ability to ride a bike, to farm, swim, walk, love etc. Functionings are therefore ways of being in the world that have been achieved by way of individuals having been free to engage in the prior realisation of capabilities, which are therefore the real possibility of achieving a certain functioning. While functionings are the means to conceptualise interpersonal comparisons of wellbeing that *have been achieved*, capabilities are the means to conceptualise interpersonal comparisons of the *freedom to pursue wellbeing*, what Sen calls 'wellbeing freedom'. Capabilities are then people's effective opportunities to undertake actions and activities that they have reason to value, and be the person that they have reason to want to be. These beings and doings together are held to constitute what makes a life valuable. A 'capability set' encompasses, according to Sen, the total functions available for a person to perform. The more limited a person's freedom is, the fewer opportunities he/she has to fulfil his/her functions.

Sen and Nussbaum both sit within the tradition of political liberalism, and this prevents them from offering a full and comprehensive theory of value. Sen does not propose a definite account of basic justice, nor a list of capabilities – he wants instead to use capabilities as a highly generalisable quality of life assessment. His approach does have a concern with justice issues, e.g. racial and gender issues. Conversely, Nussbaum defines a set of capabilities. These are Life; Bodily health; Bodily integrity (freedom of movement, from violence); Senses, imagination and thought (being able to use senses, imagine, think and reason, underwritten by the right to education); Emotions (being able to form attachments, love, be free from fear and anxiety); Practical reason (being able to form idea of 'the good' and reflexively plan one's life); Affiliation (recognise and live with others, through various forms of social interaction); Other species (i.e. concern for nature); Control over one's environment, both a) politically (e.g. participation in politics), and b) materially (e.g. property rights) (Nussbaum 2011).

The capabilities approach allows sophisticated examination of wellbeing and universal human rights by development economists and underpinned the development of a new policy paradigm called the 'human development approach'. Capabilities do not however necessarily relate to justice: the proximate causes of lack of capabilities may not have been caused by injustice, while the achievement of capabilities does not necessarily mean justice has been done. Capabilities are not parts of justice *per se*, rather they are the potentialities for wellbeing and/or individual 'freedom' or autonomy (Sen 2001).

The Analytical philosophy which underlies Capabilities has been critiqued by post-colonial scholars for its universalisation of European values and the homogenisation of diverse cultural expressions of humanity (de Sousa Santos 2015). To be sure, wellbeing and individual freedom are important components of rights and justice, but, critically, they cannot be reduced to capabilities alone. Hence the need to draw on other approaches.

Recognition

The concept of recognition originated with Hegel, and has been engaged with in different ways over the past few decades by philosophers such as Charles Taylor (1994), Paul Ricœur (2005), Nancy Fraser (1995) and Axel Honneth (1996). Charles Taylor is largely responsible for the current revival of interest in recognition with his 1994 paper 'Multiculturalism and the Politics of Recognition', which popularised the concept across social and political sciences. Taylor observed that recognition is a feature of a variety of forms of contemporary politics, including feminism, racial and gay movements and multiculturalism. Recognition here relates to identity, which is in part, 'shaped by recognition or its absence', and therefore 'Nonrecognition or misrecognition can inflict harm, can be a form of oppression, imprisoning someone in a false, distorted, and reduced mode of being' (1994: 25). Taylor emphasises that recognition is 'a vital human need' (ibid: 26) and stating that misrecognition 'can inflict a grievous wound, saddling its victims with a crippling self-hatred' (ibid: 26). Taylor's conception of recognition rests on the Hegelian notion that the individual is an inter-subjectively formed being – identity is not, as Descartes, Kant, Hobbes and Locke would hold, developed by the self alone. Instead, identity is constructed through a dialogic process with significant others i.e. family, friends, mentors, colleagues, and so on. Conversely, if I am misrecognised by those around me and broader institutions my potential self-realisation and hence freedom and wellbeing will be negatively affected.

Taking a recognition approach to rights and justice is based on the claim that human beings require specific types of relationships if they are to flourish as persons, or even become persons in the first place. The kinds of relationships they require are those of reciprocal recognition and so the provisioning of these relationships is a crucial aspect of justice. As Honneth puts it 'The justice or wellbeing of a society is proportionate to its ability to secure conditions of mutual recognition under which personal identity-formation, can proceed adequately' (Fraser and Honneth 2003: 174). Freedom here is then a social freedom which can only be made manifest, and sustained relations of mutual recognition upheld, by social institutions (Honneth 2014). Social freedom is expressed in the phrase 'being-with-oneself-in-the-other', which refers to the way in which the quality of social relations in existing practices and institutions is an essential precondition for the realisation of freedom. To self-realise, we need recognition and positive feedback from others and society more broadly throughout our lives – I know that I am a self because I see you recognising me as a self.

Political theories of recognition hold that recognition is key in determining what is just in a society and what is a good society (Thompson 2006). So recognition is important not only for the development of self-consciousness and sociality – it is also politically significant. The idea is that if we need recognition to achieve a good life, its absence drives political struggles. Disrespect and misrecognition can truly cause us harm and thus avoiding them is seen as the force behind social movements and contemporary forms of identity politics. This is where the two Hegelian ideas meet – recognition in itself is important for the good life, and hence recognition as a political concept matters when we decide upon the form a just society is to take (Hirvonen and Laitinen 2016).

Unlike more abstract and universalised analytic approaches, recognitional justice, as a form of critical theory, must arise from and be realised through an analysis of actual social institutions (Zurn 2015). This is consistent with the ethnographic tradition long employed in anthropology, human geography, development studies and sociology, all of which inform contested agronomy. The difference is that while these disciplines are primarily concerned with the description and historically informed explanation of society, recognition and other approaches within critical theory, in addition to this, are oriented to the purpose of improving human freedom and wellbeing by overcoming forms of injustice. Recognition is a form of justice as freedom, as Hegel argues; for individuals to be free (to be able to self-realise) they need recognition from the wider society, that is, intersubjective affirmation of their difference, and this can be seen as relating to both the local knowledge, and participation agendas which form part of the intellectual heritage of contested agronomy. The academic importance and impact of this body of work has been significant, despite the arguably relatively limited success of each (Cooke and Kothari 2001; Sillitoe 2010). It is possible that the injection of a more normative dimension could revitalise these agendas.

Recognition has been criticised for its role in reproducing the structural and psycho-affective dimensions of racialised (post-)colonial power relations by getting subaltern peoples to identify with asymmetrical and non-reciprocal forms of recognition either imposed or granted to them by hegemonic groups within society. Rather than remaining dependent on oppressors' 'recognition' for freedom and self-worth, subaltern peoples must instead struggle to overcome alienation/ subjection against the objectifying gaze and assimilative lure of post-colonial recognition (Coulthard 2014; Fanon 1967). According to Fanon, it is through a self-initiated process that people decolonise by first recognising themselves as free, dignified and distinct contributors to humanity. This brings us to cognitive justice.

Cognitive justice

Drawing on a 'cognitive justice' approach (Visvanathan 2005; de Sousa Santos 2007) allows us to embrace the plurality of knowledge(s) at play in any given agrarian context (Coolsaet 2016). The concept of cognitive justice is based on the acknowledgement of knowledge plurality and the rights of different forms of knowledge to co-exist equally. In the context of development-oriented agronomy it means giving equal weight to subaltern knowledges as is given to dominant positivist science. Cognitive justice therefore works as a critique of the dominant paradigm of modern science, advocating the acknowledgement of alternative paradigms or alternative sciences. It seeks to do this by facilitating and enabling dialogue between what are often incommensurable knowledges. These dialogues between knowledge(s) are perceived as contributing to a more sustainable, equitable, and democratic world. This resonates with the work on 'local knowledge' in development studies and anthropology, both that in some cases local knowledge had broadly got it 'right' in terms of sustainable natural resource management, whereas science could be shown to be 'wrong'; but also in terms of the conflicts between different knowledge(s) at play in the development field, captured in the idea of 'battlefields of knowledge' (Richards 1985; Long and Long 1992; Scoones and Thompson 1994; Fairhead and Leach 1996).

The term cognitive justice was first used by Indian scholar Shiv Visvanathan in his book *A Carnival for Science: Essays on Science, Technology and Development* (Visvanathan 1997). He argued that the hegemonic nature of Western science had a negative impact on developing countries and their non-Western cultures. He called for the recognition of alternative sciences or non-Western forms of knowledge, which underpin different livelihoods and lifestyles and therefore deserve to be treated equally. Hence, taking a cognitive justice approach means that subaltern knowledges have to be included in epistemological discourses on their own terms, that is, without being reduced to or fitted into Western knowledge frameworks. When subaltern knowledge(s) are treated equally, they can play their role in making a more democratic and dialogical science, which remains connected to the livelihoods and survival of all cultures. However, in the context of the argument advanced by this chapter, this means taking normative dimensions of subaltern knowledge more seriously.

Cognitive justice builds upon post-colonial critiques of Said, Fanon and Césare in the 1950s and 1960s which highlighted the colonial imposition of the Western development model and effects on the subjectivities of colonised peoples. The point of departure here is that understandings, ways of knowing and being in the world exceed those provided by Eurocentric analytic and continental traditions. Hence, Western-centred universal abstract human rights cannot attend to the plurality of cultural forms of respect, dignity and justice at play in the ethical life of any given social context. Indeed, the universalisation of one notion of rights forecloses the possibility of alternate lexicologies of rights and justice that may not be compatible with universal human rights. What this means is that continental philosophy, while attentive to difference and particularity, has nevertheless failed to adequately acknowledge various social actors, movements and forms of resistance in the Global South and the specific cultural, symbolic and linguistic universes they occupy.

I argue that one reason 'local knowledge' had limited success as a research agenda (Sillitoe 2010), is that not enough attention was given to the normative dimensions of that knowledge. A corollary of this therefore is that social justice cannot be achieved globally without cognitive justice. This means scholars and development practitioners, including agronomists from the Global North, learn from the South via processes of horizontal intercultural dialogue and 'translation' among different critical knowledges and practices.

In practice, cognitive justice revolves around the 'intercultural translation' of 'ecologies of knowledges' – that is, the inter-fertilisation of knowledges on an equal footing, transforming prior hegemon/subaltern knowledge dichotomies (de Sousa Santos 2015). This should be thought of less as an academic exercise than one which unfolds through the interaction of different regional and global political movements, thus entailing a plurality of knowledges, practices and agents enacting struggles against capitalism, colonialism, and patriarchy. It involves movement leaders, activists and 'subaltern intellectuals'. One example of this is the ways in which the labour movement, in its late capitalist moment of crisis, is engaging which other social movements, such as feminists, environmentalists and human rights movements, and through which there has occurred intercultural translation of practices of labour, differential claims for different kinds of (eco-)citizenship protection and demands to respect women and marginalised ethnic groups and migrants. In the context of development-oriented agronomy this would mean, for example, deeper engagement in dialogical learning with the global peasant movement Via Campesina.

Conclusion: capabilities, recognition and cognitive justice in development-oriented agronomy

The intention in this chapter has been to argue first, that consideration of rights and justice should have a central place in scholarship around knowledge politics in agronomy; and second, that capabilities, recognition and cognitive justice can provide development-oriented agronomists with a broad framework of possible intellectual tools upon which to draw. However, given the differences between analytic, continental and post-/de-colonial philosophies underlying these approaches – they are real and should be respected – it is better to hold them in productive tension vis-à-vis one another and/or use them to look at different aspects of agronomic interventions, rather than trying to integrate them.

These different approaches to rights and justice can certainly complement one another (e.g. Martin *et al.* 2016). Capabilities can be used either before or after interventions to explore the ways in which the capabilities of different individuals are either realised or impinged upon (or indeed unaffected by) interventions. In one sense, recognition has already entered the realm of contested agronomy in the struggles around both local knowledge and participation. The act of studying local knowledge and attempts to foster participation are valuable, no doubt flawed as they are, and they can be seen as attempts at certain kinds of recognition. The contested agronomy research agenda includes politics and power, and recognition is useful to examine experiences of subordination and domination by subaltern groups often targeted by agricultural development projects. However, we need to address post-colonial critiques of recognition, by radically questioning whether politically available forms of recognition (e.g. rights) really enable self-realisation or are new forms of domination. Here, cognitive justice can help by engaging with social movements in processes of dialogical learning around agricultural interventions (e.g. Meek and Simonian 2016). Cognitive justice is essential in order to question whether what emulates from Western models (whether individualist or communitarian) are the only pathways to development. This undermines long-held assumptions associated with modernisation theory by valorising knowledge and realities of marginalised and subaltern populations of farmers and their taking the goal of their participation in agricultural development seriously. There are manifold emancipatory discourses and practices in the Global South which do not fit with Western assumptions – not just ethical, political, cultural, but also epistemological and ontological – which are rendered invisible by Western development models. De Sousa Santos (2007) argues that understanding this diversity is critically important.

In conclusion, this framework and the complementarity of the different approaches it is built upon, can be put to work as follows. Capabilities permit the conceptualisation of freedom or flourishing for individual people in the contexts of agricultural development projects (highlighting agency). Recognition allows us to reveal current and past injustices suffered by individuals and groups embedded in social and institutional contexts (highlighting structure). And cognitive justice gives us the tools to appreciate subaltern epistemologies and ontologies at play in any given situation (highlighting knowledge pluralism). Because such 'theory from the south' is necessarily contextual and often non-universalistic, it seems logical that the cognitive justice dimension draws on whichever local or regional philosophical traditions are appropriate to a particular context or intervention.

Drawing creatively on this framework, development-oriented agronomists can make an explicit examination of the likely rights and justice effects of new agronomic technologies, either at the planning stage, and/or in evaluations of impacts on rights and justice following implementation or promotion.

Acknowledgements

I would like to thank colleagues at Lancaster and UEA for the stimulating conversations upon which this chapter leans heavily. They are Saskia Vermeylen, Ben Neimark, Brendan Coolsaet, Iokiñe Rodríguez, Adrian Martin, Gareth Edwards, Esteve Corbera, Neil Dawson and Ina Lehman. I also would like to thank participants of the conference 'Contested Agronomy: Whose Agronomy Counts?', 23–25 Feb 2016 at the Institute of Development Studies, for comments on an earlier version of this chapter which contributed to its improvement. All of the inevitable errors of philosophical understanding rest with the author, however.

Notes

1   http://www.futureearth.org/
2   http://www.unac.org.mz/index.php/artigos/nacional/94-campanha-nao-ao-prosavana-mocambicanos-pedem-solidariedade-regional
3   http://histphil.org/2016/01/04/was-the-green-revolution-a-humanitarian-undertaking/
4   e.g. http://www.comboni.org/pt/contenuti/107453-mo-ambique-n-o-aos-a-ambarcadores-de-terras; http://www.unac.org.mz/index.php/artigos/nacional/129-outra-vez-o-prosavana; http://www.unac.org.mz/index.php/artigos/nacional/94-campanha-nao-ao-prosavana-mocambicanos-pedem-solidariedade-regional

11

A GOLDEN AGE FOR AGRONOMY?

*Ken E. Giller, Jens A. Andersson, James Sumberg
and John Thompson*

Introduction

Agriculture and food are back on the global agenda, and as a result, agronomic research is experiencing something of a renaissance. However, as argued in *Contested Agronomy: Agricultural Research in a Changing World* (Sumberg and Thompson 2012) and illustrated by the chapters in this book, this is a renaissance with a difference. The new agronomy spills beyond the traditional bounds of the discipline, not least because the environment in which it takes place has dramatically altered. Changes in funders and their funding priorities, the introduction of New Public Management (NPM) principles, a singular focus on *impact,* and an altered position of science in society, characterise this changing environment (Duval *et al.* 2015; Renkow and Byerlee 2009). As a result, the new agronomy is both more public and more contentious than ever before.

The different ways that agronomy's purpose, priorities and success are framed by agronomists, research funders, policy makers, lobby groups, and others are key to understanding knowledge politics in contemporary development-oriented agronomy. As the cases in this book have highlighted, these politics really matter. They are reshaping the discipline of agronomy and its direction, and in so doing, the ability of development-oriented agronomy to support rural livelihoods and more productive and sustainable food systems.

In this final chapter, we further explore knowledge politics in development-oriented agronomy. In particular, we focus on dominant framings and popular research themes: the rise of the new global food security agenda, which marks agronomy's renaissance and reframing as a global food security science; the framing and reframing of long-standing debates related to farm size and agronomic practices; questions over scales of analysis and the evolution from plot-based to place-based agronomy; and the growing influence of the New Public Management principles on agricultural research priorities and processes. We suggest that development-oriented agronomy is at a critical juncture, facing a once-in-a-generation opportunity to re-imagine itself.

Global food security: agronomy's new princess?

For most of the twentieth century, agronomy was seen as a technical discipline which focused on the important, but largely practical matter of improving crop production. In this context, but particularly following the growth of state planning and economic management in the 1930s, politics was mainly limited to interactions between the state, which set agricultural policy priorities and funded research, and national research organisations and universities, where agronomists undertook their research and in turn informed agricultural and development practice. Agronomic research played the role of handmaiden to the state. The rise of neoliberal policies and the emergence of the environmental and participation movements in the 1970s undermined that comfortable relationship and the unity of purpose on which it was built (Sumberg, Thompson and Woodhouse 2012). Together, these forces fostered a political climate in which state-led agendas for science were increasingly questioned. New actors, interests and value orientations entered the scene, both from the private sector and from civil society. They began to challenge the 'normal science' of agronomy, questioning long-established norms and practices, and reorienting and redefining the priorities of agronomic research. In recent years, these contestations have ratcheted up because the world's agri-food systems are facing increasingly dynamic, uncertain and complex ecological, social and technological challenges (Thompson and Scoones 2009).

The food price spikes of 2007–2008 (von Braun 2011) can be seen as a marker of another realignment of actors, interests and value orientations around development-oriented agronomy. Volatile food prices, an extra billion mouths to feed by 2050, and accelerating climate change – the so-called 'perfect storm' – brought the gaze of politicians and other influential people back to agriculture and agricultural research. The agricultural research community responded to this new sense of crisis around global food security by re-asserting the ability of science and technology to deliver solutions: *'Yes, with "the best" science, we can feed the world!'*

Agricultural research was reframed as 'Global Food Security' research. For instance, the US Government established a new flagship programme, Feed the Future, for investment in global food security. The UK Government also launched a new Global Food Security Programme to meet the challenge of providing the world's growing population with a sustainable, secure supply of safe and high-quality food using less inputs, and in the context of global climate change. The Australians followed suit, establishing a Food Security Centre in 2012 (but closing it down again in 2015).

Also at the multilateral level, investments and initiatives abound. The World Bank launched its US$ 1.6 billion Global Agriculture and Food Security Program (GAFSP) following the food price spikes. The UN Food and Agriculture

Organization (FAO) aimed to better capture key aspects of the renewed focus on food security launching a new set of indicators in their annual publication *State of Food Insecurity in the World* (FAO 2012, 2013b, 2014, 2015, 2016). The UN's 17 Sustainable Development Goals (SDGs), launched in 2014, also reflect this renewed focus; SDG 2 seeks to end hunger, achieve food security and improved nutrition, and promote sustainable agriculture by the year 2030 (UN 2015).

The renewed focus on global food security has not merely meant increased investment in agricultural development, but has been accompanied by a boost in the funding of agricultural research. The accelerated growth of international agricultural research through the Consultative Group on International Agricultural Research (CGIAR) is illustrative. From an annual growth of about 5 per cent over the period 1998–2007, contributions to the CGIAR increased to an average of 12 per cent annually over the period 2008–2014 (CGIAR 2012, 2014). Yet, in the plethora of initiatives and new investments around global food security, there is much jockeying among private, bilateral and multilateral organisations as they attempt to influence agricultural research, with a strong drive toward research for impact. This progressive blurring of the boundaries between research and development has injected new energy into knowledge politics around agricultural research, and particularly through the framing of agronomic priorities to which we turn now.

Green revolutions and yield gaps

It is important to note that despite dramatic changes in the geopolitical context, in economic orthodoxy, in biological science and in the diversity and interests of funders, the central imaginary for the new assault on global food security was plucked from the past – the Green Revolution. Why? Because this imaginary feeds a belief that the central challenge is to get the science and technology right. All the messy and uncomfortable truths about institutions, politics, power, and winners and losers, can be set aside. This allows a focus on the 'availability' pillar of food security with scant acknowledgment of, or attention to 'access' or 'distribution' issues, let alone environmental costs. This is nowhere more evident than in the now nearly hegemonic emphasis on closing 'yield gaps'. While breeders focus on increasing the genetic potential of crop varieties (and widening the yield gap), agronomists focus on management systems that move genetic potential to realised yield. This is the pinnacle of agronomy as engineering.

The yield gap is an extremely powerful mobilising device. Simply declaring a gap acts as a call to arms for it to be filled. It is certainly true that a better understanding of where and why yield gaps occur can allow considered reflection on the reasons why yields are so poor and yield gaps so large in some countries and regions, and why yields are so large and yield gaps small in others (van Ittersum *et al.* 2013). But what was once simply a core concept in crop ecology has been hi-jacked in policy circles (and in agronomists' research proposals) to highlight the need for increased productivity (Sumberg 2012). Most references to the yield gap do not indicate any appreciation of the context specificity of yield gaps that is so central to their treatment in crop ecology. Rather the notion of yield gaps is used as if they are absolute and universal. We run the risk that the yield gap framing of the problems of food and agriculture reduces consideration to only technical solutions, while anything to do with markets, institutions and livelihoods goes out the door. Technology transfer – and adoption – are again at centre stage, and all that has been learned about processes and dynamics of technological change has been jettisoned. This is certainly a retrograde step.

Yet all is not lost! The on-going work on yield gaps is helping to identify a plethora of gaps – efficiency gaps, technology gaps, nutrient gaps and gaps in ecosystem services. This research is opening up new understanding of the underlying causes of yield gaps at different levels beyond the plot, and a recognition that a narrow focus on improving efficiency alone will not achieve the desired results (e.g. Van Noordwijk and Brussaard 2014; Silva *et al.* 2016; Stuart *et al.* 2016). These ideas are explored further in a forthcoming special issue of the journal *Experimental Agriculture* focusing on central concepts and research methods in contemporary development-oriented agronomy.

The framing of farming: is simplification illuminating?

Small or large farms, family or corporate farms

Agriculture and farming are framed in many ways, and for many purposes. The United Nations General Assembly declared 2014 the *International Year of Family Farming*, clearly equating family farms with small farms (FAO 2014b). The overall argument is that family and small-scale farming are inextricably linked to world food security, as they are reportedly responsible for producing more than 70 per cent of the world's food supply. In low-income countries, farms smaller than two hectares are estimated to produce more than 40 per cent of the food and farms below five hectares more than 70 per cent (Lowder *et al.* 2014). Yet this view of family farming is problematic. For example, if family farms are defined by their two most distinguishing attributes – namely inheritance of the farm and the predominant use of family labour – then family farms range in size from much less than one hectare to more than 10,000 ha (van Vliet *et al.* 2015). Using this definition, almost all farms throughout the world are family farms, yet they encompass a huge diversity of production methods and relations.

In both academic and policy circles there seems to be an ineluctable compulsion to romanticise small farms and smallholder farmers. Farming is no doubt a way of life that many choose, yet in developing countries people are often farming by default. Many consider themselves to be unemployed. A stark finding from analysis of more than 13,000 farm surveys across 393 sites in 17 countries of sub-Saharan Africa was that almost 40 per cent of responding households were unable to achieve food self-sufficiency, and those that did were dependent on off-farm income (Frelat *et al.* 2016). Indeed, small may be 'beautiful', as E. F. Schumacher famously asserted (Schumacher 1973), but how small (Hengsdijk *et al.* 2014)?

Family farms and the implicit values and practices they represent are frequently contrasted with corporate or industrial agriculture. But despite being used in a pejorative way, industrial agriculture is rarely if ever defined. The implication is that it is large-scale, controlled by corporate interests, focused on quantity not quality, environmentally destructive and blind to animal and human welfare concerns. Industrial agriculture is thus portrayed as the antithesis of smallholder agriculture and agro-ecological approaches. These kinds of dichotomous framings have been used to justify calls for the 're-peasantisation of agriculture' (van der Ploeg 2010).

Organic or conventional farming

There are other examples of simplified framings of farming that are more mystifying than illuminating. We have a reasonable idea of what constitutes organic farming, as this is set out by legal standards (Willer and Kilcher 2016). But organic farming is often compared and contrasted with conventional farming. Everyone has an idea of what conventional farming is but, oddly, it is never defined. If conventional farming is defined simply by not being organic, then pretty much the whole diversity of farming globally is lumped into one category. The same applies to alternative agriculture – alternative to what? Are these terms simply convenient devices used to argue a specific case?

The example of agroecology as the sustainable alternative to industrial agriculture is worth critical consideration. Promoters of agroecology often point to a lack of research funding and suggest that if more resources had been given to development of agroecological approaches then these would be better understood and more economically viable. This is a plausible argument, but it is also highly problematic in that all forms of agricultural research, including agroecology, build on a rich tradition of biological knowledge. Molecular biology has undoubtedly devoured the majority of research funding in the biological sciences over the past decades. Yet the critique by Vanloqueren and Baret (2009) that this has all been funnelled toward the development of genetically engineered crops misses the point. Much of the research has been directed to elucidating fundamental biological processes. The same applies for much of the research conducted in other fields such as soil biology and nutrient cycling, including the work of the Tropical Soil Biology and Fertility Institute of the International Center for Tropical Agriculture (TSBF-CIAT) which gave ecologists a central place in agricultural research (Woomer and Swift 1994; Cadisch and Giller 1997; Van Noordwijk et al. 2004). Although this research was not framed as organic, alternative or agroecological, and did not specifically target chemical-free forms of production, it provides a firm basis for our understanding of processes that contribute to agroecology.

Perhaps more provocatively, if we consider that agroecology is fundamentally about harnessing biology and biological processes, then it could be argued that genetic engineering is one of the most powerful (potential) tools of agroecology (see Box 11.1).

BOX 11.1 TRIPLE-STACKED ARGUMENTS FOR AND AGAINST GM CROPS

The potato famine that occurred in Ireland in the mid-nineteenth century was triggered by massive crop loss resulting from infestation with late blight (*Phytophthora infestans*) although politics played a key role in magnifying the consequences. Late blight remains a major problem for potato production worldwide, not least because the plasticity of the fungal pathogen means that genetic resistance breaks down rapidly. In 2006, the 'DuRPh' project (Sustainable Resistance to *Phytophthora*) project was established to develop triple-stacked resistance (Haverkort *et al.* 2008). The project used CIS-genetics to transfer resistance genes from natural relatives of potato. These genes could have been transferred by traditional breeding methods although this would take several decades to achieve. The resulting varieties require a single spray of fungicide per season as opposed to the 15 or more sprays required with other commercial varieties. With new gene splicing methods such as CRISPR-CAS – such genetic changes can be made without leaving a trace. Integrated approaches to disease control using such resistant varieties together with crop rotation could reduce the need for fungicide to control late blight drastically. The choice for society – and for those promoting alternative agro-ecological approaches – is whether to use this GM approach and save thousands of litres of fungicide each year or continue to exclude the technology.

Conversely, the case for GM crops with stacked herbicide resistance genes under discussion in the US, which will inevitably lead to much greater use of herbicides, is exceedingly difficult to justify on any agroecological grounds. This illustrates the weaknesses of the dichotomous framings that shape much public and policy debate around agriculture, and the need for careful and nuanced consideration of potential GM benefits to society and the environment on a trait-by-trait basis.

An idea often pursued is that agricultural systems have been oversimplified and should mimic natural systems that are more biodiverse and structurally complex. Yet, given that farming is designed to produce and export food, there is no *a priori* reason why the organisation of natural systems should be superior to well-designed agricultural systems (Denison and McGuire 2015). Wood and Lenné (2001) argued that highly stable monocultures are common in nature and could be an equally useful model for study when considering the attributes of sustainable ecosystems that are relevant to farming.

So what do we conclude? First, that dogma has no place in agronomic science. Second, that agronomists must become more aware of the framing of agronomic problems and solutions; that knowledge production is a value-directed process, and consequently, that agronomic knowledge has to be understood within a particular context. So rather than isolating ourselves by focusing on (universal) technical solutions, agronomists need to embrace agronomy as a situated, place-based science, and to do this they must engage more openly with the other agricultural and social disciplines.

From plot-based to place-based agronomy

The basic principles of agronomy – including production ecology and the crop and soil sciences – are now well understood. We know that crops respond to applied nitrogen when it is the factor in limiting supply; that irrigation prevents the impact of drought on crop growth; that drainage limits the effects of waterlogging; and that pests and diseases must be controlled. Our understanding of the interactions among the various limiting and reducing factors is also fairly well-developed. But for agronomy to be relevant we have to understand how things work in the real world of farming; not only under controlled experimental conditions.

Given the huge diversity of farms described above, it is inevitable that there are multiple plausible answers to any specific question raised. A place-based agronomy seeks to understand how the basic principles of agronomy are conditioned by the range of agroecological and socioeconomic circumstances within which farming takes place. It is more concerned with understanding the risk of success or failure of a technology (Vanlauwe *et al.* 2016), and how the technology fits for different farmers (Giller *et al.* 2011) than with average responses.

Returning to knowledge politics, the saying 'all politics is local' is common in the United States and it encapsulates the principle that a politician's success is directly tied to his/her ability to understand and influence the issues important to local constituents. In the context of development-oriented agronomy, the variant 'all agronomy is local' is equally compelling. The real challenge is to bridge the gap between, on the one hand, the valuable insight that is at the heart of this saying, and on the other, the expectations of 'impact at scale', and the increasingly ubiquitous framing of 'grand global challenges'. This brings into focus the interplay between principles and place-based practice, with the critical point being that 'place' implies not only an agroecological setting (soil, water, climate, etc.), but also the social, institutional, economic, political, historical and cultural contexts within which crops are grown and farming is performed. We question whether the field of agronomy has been creative enough in thinking through the continuum from 'principles to place and practice'. There is a need for a new 'systems agronomy' (Giller *et al.* 2015) that works at and across the levels of the farming (household) system, the agricultural system and beyond.

Framing development-oriented agronomy: a New Public Management straightjacket?

It is difficult for anyone to argue against greater transparency and accountability, particularly in an age of austerity when there are growing demands on the public purse. For agricultural research, the question is more about what transparency and accountability mean, and how they can best be achieved.

Agronomic research has two distinct but interwoven strands. One strand asks 'why?', while the other asks 'how?'. It is the interplay of 'why?' and 'how?' within agronomy that has spawned the knowledge, technological breakthroughs and institutional innovations that underpin the agricultural productivity gains of the last 100 years. Development-oriented agronomy (and agricultural research for development more broadly), despite an explicit problem-solving orientation, cannot afford to focus only on the 'how?' questions.

Indeed, agricultural research is expected to play its part in addressing the global grand challenges, including global food security, climate change adaptation, de-carbonisation and so on. It is commonly asserted that these challenges will require vision, innovation, interdisciplinarity, thinking 'outside the box', risk taking etc. This points to an important truth: addressing either 'why?' or 'how?' questions *always* involves a foray into the unknown, where outcomes are uncertain, definitions of success are ambiguous, and occasional failure cannot be avoided (and indeed may provide extremely valuable opportunities to learn). It is against this background that responses to the call for greater transparency and accountability have come to cast an increasingly long shadow over development-oriented agronomy.

New Public Management provided the ideological and intellectual underpinnings for reform of the public sector so that it would be more transparent and accountable, and above all, more efficient. In agricultural research the principles of NPM are reflected in the central place that results frameworks, impact pathways, theories of change and results-based management now play in programme and project planning, management, reporting and evaluation. In effect, researchers are asked to specify the outcomes and impacts of their work long before it is undertaken. In this sense the mentality that goes along with results frameworks and results-based management is diametrically opposed to discovery, innovation, creativity and learning from failure, which, with or without a 'development orientation', are at the very heart of the research, knowledge and technology generation enterprise that is agronomy. As such we believe that this mentality has important implications for the practice of development-oriented agronomy and its potential contribution as a science. Specifically it is fuelling a tendency for present-day development-oriented agronomy to veer to the pragmatic and short-term, where the extent of the enquiry is limited to whether or not something 'works'. At the moment there is no empirical evidence about the breadth or depth of this tendency, and this now deserves serious research attention.

Even the most successful research agronomists have always been under some pressure to justify their funding, and presumably most have thought their funding was chronically inadequate. Nevertheless, the disciplining tools of NPM have completely changed the agricultural research landscape, and the intellectual landscape of agronomy, and would be seen as alien (if not an anathema) by agronomists of only a generation ago, including the 'father' of the Green

Revolution (Box 11.2). How can it make sense to train and employ scientists and then send them out to explore the unknown with both hands tied behind their backs?

BOX 11.2 AN IMAGINED CONVERSATION

Funder: So, Dr Borlaug, tell us again what you are planning to do in Mexico?

Norman Borlaug: I am planning to work hard and stay focused, and produce rust resistant wheat varieties.

Funder: Yes, yes, we know all that, but what is your *theory of change*?

NB: Theory of change? *What?* Didn't you take Agriculture 101 at university? Oh, I see, well, let me spell it out for you: disease free crops are more productive than disease ravaged crops, and therefore the availability of disease resistant crop varieties is a keystone in the fight against hunger.

Funder: But Dr B. that just won't do! A theory of change must specify Immediate Outcomes, Intermediate Outcomes, Impacts, Assumptions, Contextual Factors, Rival Explanations, Unintended Results; and, most important of all, the relationships between all of these must be clearly displayed in a compelling diagram. Can you please work on this and get back to us in a couple of days. Oh, and while you are at it, please be sure that you fill out the section of the grant application form on Pathways to Impact – it was probably a clerical error, but you seem to have left it blank in your initial submission.

NB: Clerical error my foot! My job is to be in the field, making crosses, planting and walking the nurseries, scoring lines for rust resistance, making selections, training staff and the next generation of scientists, running trials and talking to farmers. That is the only impact pathway that matters. Are you seriously suggesting that it would be better if I were to use my time fabricating theories of change and impact pathways?

Funder: Yes, now run along and get on with it – and please don't forget to specify exactly how our funding of your work will lead to impact at scale!

If development-oriented agronomy is to make its contribution to addressing the global grand challenges, a new balance needs to be struck between short and long term, pragmatism and vision, assurance and risk, and control and creativity. This is not about agronomists wanting to cloister themselves in an ivory tower, work on interesting but irrelevant problems, or simply writing papers to further their careers. Rather, it is about allowing them the freedom, space and resources to do what they are trained and eager to do. If the dead hand of NPM and results-based (research) management is not removed from the throat of development-oriented agronomy, it is a near certainty that the grand challenges around agriculture and food will be with us for many decades to come.

A golden age of (transformational) agronomy?

Agronomy is increasingly operating in a space where uncertainty is large and the stakes are high, and where many opinions and ideas are heard – what Ravetz (1999) called 'post-normal science'. In this new world, questions have more than one plausible answer, and quality and rigour take precedence over truth or factual knowledge. In post-normal science, knowledge politics take central stage and is unavoidable. Yet, in recent examples of contested agronomy, perhaps bolstered by the need to 'make one's case', there has been a tendency to take selectively from, de-value or even reject academic knowledge. Rather than trying to engage in reasoned debate, testimonies of selected farmers have been used to de-bunk scientific analysis.[1] Given the plurality of legitimate perspectives about agriculture, and our increasingly interconnected world, such exclusionary approaches will do little to advance knowledge or resolve contestation.

This is not to suggest that knowledge politics will, in the end, have only limited influence on the process of agronomic knowledge production. The growing importance of NPM principles in the management of agricultural research, in the form of theories of change, results frameworks, impact pathways, and (falsification-denying) validation trials, threaten to put development-oriented agronomy into a pragmatist straightjacket in which research is merely evaluated in terms is of its direct practical use and impact.

For agronomy to live up to its promise and contribute to the global challenges around food and farming facing humanity, we need to learn to live with – and embrace – the politics of knowledge. Rather than behaving as alchemists – peddling 'false gold' in the form of universal solutions to situated problems – agronomists must tackle the diversity and complexity of agriculture and agronomy head on. Surely this is where the interesting science is yet to be done.

Agronomists often look jealously at (more easily measured) impacts, such as the uptake of genetic resources developed by crop breeders. Rather than blaming field-scale agronomy for not delivering impact at scale, we need to understand the (often dysfunctional) context within which farming takes place, the lack of take-up of technology due to its poor fit to farmers' circumstances, and the barriers that farmers may face in accessing technology. Armed with this knowledge we can explore alternative policies and pathways that can enable the generation of agronomic approaches tailored to farmers' needs and aspirations.

Finally we return to the food price crisis of 2008, and to everyone's relief commodity prices have since fallen back toward pre-crisis levels. If this trend continues, and the perfect storm appears to pass, global attention to agriculture and food security may again wane. If so, will we blink and miss the chance to claim a Golden Age for Agronomy?

Acknowledgment

Research for this paper was partially funded by CGIAR Research Programs on MAIZE and WHEAT.

Note

1   The Howard G. Buffett Foundation 'You can listen to the academics or you can listen to the farmers', http://harvestingthepotential.org/brownrevolution/assets/doc_01.pdf (accessed 1 November 2016).

REFERENCES

AATF (2010) *Rationale for a Biosafety Law for Uganda*, WEMA Brief, Nairobi: African Agricultural Technology Foundation.

Abdenur, A. E. and da Fonseca, J. M. E. M. (2013) 'The north's growing role in south–south cooperation: Keeping the foothold', *Third World Quarterly*, 34(8): 147–1491.

ACB (2015) *Nuanced Rhetoric and the Path to Poverty: AGRA, Small-scale Farmers, and Seed and Soil Fertility in Tanzania*, Melwille, South Africa: African Centre for Biosafety.

Acemoglu, D. and Robinson, J. (2012) *Why Nations Fail: The Origins of Power, Prosperity, and Poverty*, New York: Crown Publishers.

Adenle, A. A., Morris, E. J. and Parayil, G. (2013) 'Status of development, regulation and adoption of GM agriculture in Africa: Views and positions of stakeholder groups', *Food Policy*, 43: 159–166.

Adger, W. N. (2000) 'Social and ecological resilience: are they related?', *Progress in Human Geography*, 24: 347–364.

Adipala, E., Laker, C., Tizikara, C. and Wadhawan, H. (2003) *NAADS Second Joint Government of Uganda-Donor Review, Final report* (April–May 2003), Kampala.

AFSA (2014) Appeal to ARIPO and African Union member states and UNECA for an effective aripo protocol supportive of farmers' rights and the right to food. Harare, Zimbabwe.

Akinnagbe, O. M. and Ajayi A. R. (2010) 'Challenges of farmer-led extension approaches in Nigeria', *World Journal of Agricultural Sciences*, 6(4): 353–359.

Akram-Lodhi, A. H. (2013) *Hungry for Change: Farmers, Food Justice and the Agrarian Question*, Halifax: Fernwood Publishing.

Akrich, M. (1992) 'The de-scription of technical objects', in Bijker, W. E. and Law, J. (eds) *Shaping Technology/ Building Society: Studies in Sociotechnical Change*, Cambridge, MA: MIT Press, pp. 205–224.

Alden, C., Morphet, S. and Vieira, M. A. (2010) *The South in World Politics*, London: Palgrave Macmillan.

Allan, W. (1965) *The African Husbandman*, Edinburgh: Oliver and Boyd.

Alliance for a Green Revolution in Africa (AGRA) (2014) *Planting the Seeds of a Green Revolution in Africa*, Nairobi: AGRA.

Almekinders, C. J. M., Louwaars, N. P. and Debruijn, G. H. (1994) 'Local seed systems and their importance for an improved seed supply in developing countries', *Euphytica*, 78: 207–216.

Alvarez, S., Douthwaite, B., Thiele, G., Mackay, R., Cordoba, D. and Tehelen, K. (2010) 'Participatory impact pathways analysis: A practical method for project planning and evaluation', *Development in Practice*, 20(8): 946–958.

Amanor, K. S. (1999) *Global Restructuring and Land Rights in Ghana: Forest Food Chains, Timber and Rural Livelihoods*, Research Report 108, Uppsala, Sweden: Nordic Institute of African Studies.

Amanor, K. S. (2010) *Participation, Commercialisation and Actor Networks: The Political Economy of Cereal Seed Production Systems in Ghana*, FAC Working Paper 16, Brighton, UK: Future Agricultures Consortium.

Amanor, K. S. (2015) *Rising Powers and Rice in Ghana: China, Brazil and Agricultural Development*, FAC Paper no 123, Brighton, UK: Future Agricultures Consortium.

Amanor, K. S. and Chichava, S. (2016) 'South–South cooperation, agribusiness, and African agricultural development: Brazil and China in Ghana and Mozambique', *World Development*, 81: 13–23.

Anan, K. (2008) 'Preface', in Yuksel, N. *Achieving a Uniquely African Green Revolution, Report and Recommendations from a highlevel conference and seminar at the Salzburg Global seminar 2008, Brighton: IDS, Salzburg Global Seminar & Future Agricultures Consortium.*

Anderson, E. (2011) *Prairie Volta Ltd. Ghana Rice Project, Public Private Partnership Presentation*, 1–2 March, Washington, DC: Prairie Volta Ltd.

Anderson, J. R. (1998) 'Selected policy issues in international agricultural research: On striving for international public goods in an era of donor fatigue', *World Development*, 26(6): 1149-1162.

Anderson, J. R. (2006) *The Rise and Fall of Training and Visit Extension: An Asian Mini-Drama with an African Epilogue*, World Bank Policy Research Working Paper 3928, Washington, DC: The World Bank.

Anderson, R. S., Levy, E. and Morrison, B. M. (1991) *Rice Science and Development Politics: Research Strategies and IRRI's Technologies Confront Asian Diversity (1950–1980)*, Oxford: Clarendon Press.

Andersson, J. A. and Giller, K. E. (2012) 'On heretics and God's blanket salesmen: Contested claims for conservation agriculture and the politics of its promotion in African smallholder farming', in Sumberg, J., Thompson, J. (eds), *Contested Agronomy: Agricultural Research in a Changing World*, London: Routledge.

Andersson, J. A. and Sumberg, J. (2017) 'Knowledge politics in development-oriented agronomy', in Sumberg, J. (ed.), *Agronomy for Development: The Politics of Knowledge in Agricultural Research*, London: Routledge.

Andersson, J. A., Giller, K. E., Sumberg, J. and Thompson, J. (2014) 'Comment on "Evaluating conservation agriculture for small-scale farmers in Sub-Saharan Africa and South Asia" [Agric. Ecosyst. Environ. 187 (2014) 1–10]', *Agriculture, Ecosystems and Environment*, 196: 21–23.

Annan, K. A. (2007) Remarks on the launch of the Alliance for a Green Revolution in Africa. World Economic Forum, Cape Town, South Africa, June 14.

Anshan, L. (2008) *China's New Policy Towards Africa*, in Rotburg, R. (ed.), *China into Africa: Trade Aid and Influence*, Washington, DC: Brookings Institution Press, pp. 21–49.

ARIPO (2013) Responses to the comments made by civil society organisations ARIPO-CM-XIV-8-ANNEX I. Lilongwe, Malawi.

Arkesteijn, M., van Mierlo, B. and Leeuwis, C. (2015) 'The need for reflexive evaluation approaches in development cooperation', *Evaluation*, 21(1): 99–115.

Armstrong, J. (2013) *Improving International Capacity Development: Bright Spots*, London: Palgrave Macmillan.

Asiedu, E. (2003) 'Debt relief and institutional reform: A focus on heavily indebted poor countries', *The Quarterly Review of Economics and Finance*, 43(4): 614–626.

Aune, J. B. and Bationo, A. (2008) 'Agricultural intensification in the Sahel – The ladder approach', *Agricultural Systems*, 98(2): 119–125.

Baker, K. M. (1985) 'The Chinese agricultural model in West Africa', *Pacific Viewpoint*, 26(1): 401–414.

Barnett, C. (2011) 'Geography and ethics: Justice unbound', *Progress in Human Geography*, 35: 246–255.

Barnett, C. (2012) 'Geography and ethics: Placing life in the space of reasons', *Progress in Human Geography*, 36: 379–388.

Barnett, C. (2014) 'Geography and ethics III: From moral geographies to geographies of worth', *Progress in Human Geography*, 38: 151–160.

Bassett, T. J. (2001) *The Peasant Cotton Revolution in West Africa. Côte d'Ivoire, 1880–1995*, London: Cambridge University Press.

Baumgaertner, B. and Holthuijzen, W. (2016) 'On nonepistemic values in conservation biology', *Conservation Biology* (doi:10.1111/cobi.12756).

Bawden, R. (1995) 'On the systems dimensions of FSR', *Journal for Farming Systems Research and Extension*, 5: 1–9.

Becker, L. (1994) 'An early experiment in the reorganisation of agricultural production in the French Soudan (Mali), 1920–40', *Africa*, 64(3): 373–390.

Beckert, S. (2014) *Empire of Cotton: A New History of Global Capitalism*, London: Penguin Books.

Behrman, J. R., Alderman, H. and Hoddinott, J. (2004) 'Nutrition and Hunger', in Lomborg, B. (ed.), *Global Crises,* Cambridge, UK: Cambridge University Press.

Beinart, W. (1984) 'Soil erosion, conservationism and ideas about development: A Southern African exploration, 1900–1960', *Journal of Southern African Studies*, 11(1): 52–83.

Beintema, N. M. and Stads, G. J. (2006) *Agricultural R&D in Sub-Saharan Africa: An Era of Stagnation,* Washington, DC: International Food Policy Research Institute.

Bello, W. (2004) *Deglobalization: Ideas for New World Economy,* London: Zed Books.

Bellon, M. and Brush, S. (1994) 'Keepers of maize in Chiapas, Mexico', *Economic Botany*, 48: 196–209.

Bellon, M. and Risopoulos, J. (2001) 'Small-scale farmers expand the benefits of improved maize germplasm: A case study from Chiapas, Mexico', *World Development*, 29: 799–811.

Bellon, M. R. and Hellin, J. (2011) 'Planting hybrids, keeping landraces: Agricultural modernization and tradition among small-scale maize farmers in Chiapas, Mexico', *World Development*, 39: 1434–1443.

Bellon, M. R., Hodson, D. and Hellin, J. (2011) 'Assessing the vulnerability of traditional maize seed systems in Mexico to climate change', *Proceedings of the National Academy of Sciences of the United States of America*, 108: 13432–13437.

Benin, S., Ephraim, N., Okecho, G., Randriamamonjy, J., Kato, E., Lubade, G. and Kyotalimye, M. (2012) 'Impact of the National Agricultural Advisory Services (NAADS) program of Uganda: Considering different levels of likely contamination with the treatment', *American Journal of Agricultural Economics*, 94(2): 386–392.

Benouniche M., Kuper M., Poncet J., Hartani T. and Hammani A. (2011) 'Quand les petites exploitations adoptent le goutte-à-goutte: initiatives locales et programmes étatiques dans le Gharb (Maroc)', *Cahiers Agricultures*, 20: 40–47.

Benouniche, M., Zwarteveen, M. and Kuper, M. (2014) 'Bricolage as innovation: Opening the black box of drip irrigation systems', *Irrigation and Drainage*, 63(5): 651–658.

Berkes, F., Folke, C. and Colding, J. (2000) *Linking Social and Ecological Systems: Management Practices and Social Mechanisms for Building Resilience*, Cambridge: Cambridge University Press.

Berkhout, E. and Glover, D. (2011) *The Evolution of the System of Rice Intensification as a Socio-technical Phenomenon: A Report to the Bill & Melinda Gates Foundation*. Wageningen, NL: Wageningen University and Research Centre.

Berkhout, E., Glover, D. and Kuyvenhoven, A. (2015) 'On-farm impact of the System of Rice Intensification (SRI): Evidence and knowledge gaps', *Agricultural Systems*, 132: 157–166.

Bernauer, T. and Meins, E. (2003) 'Technological revolution meets policy and the market: Explaining cross-national differences in agricultural biotechnology regulation', *European Journal of Political Research*, 42: 643–683.

Bernstein, H. (2016) 'Agrarian political economy and modern world capitalism: The contributions of food regime analysis', *Journal of Peasant Studies*, 43(3): 611–647.

Bettini, G., Nash, S. L. and Gioli, G. (2016) 'One step forward, two steps back? The fading contours of (in)justice in competing discourses on climate migration', *The Geographical Journal* (doi:10.1111/geoj.12192).

Biggs, S. and Smith, G. (1998) 'Contending coalitions in agricultural research and development: Challenges for planning and management', *Knowledge and Policy*, 10(4): 77–89.

Biggs, S., Matsaert, H. and Justice, S. (2016) 'Use of innovation systems analysis in agricultural and rural development: Personal reflections', Unpublished paper.

Biggs, S. D. (1990) 'A multiple source of innovation model of agricultural research and technology promotion', *World Development*, 18(11): 1481–1499.

Biggs, S. D. and Clay, E. J. (1981) 'Sources of innovation in agricultural technology', *World Development*, 9(4): 321–336.

Bingen, R. J. (1998) 'Cotton, democracy and development in Mali', *The Journal of Modern African Studies*, 20(4): 497–512.

Birner, R. and Resnick, D. (2010) 'The political economy of policies for smallholder agriculture', *World Development*, 38(10): 1442–1452

Birner, R. and Wittmer, H. (2003) 'Using social capital to create political capital – how do local communities gain political influence? A theoretical approach and empirical evidence from Thailand', in Dolšak, N. and Ostrom, E. (eds) *The Commons in the New Millennium, Challenges and Adaptation*, Cambridge, MA and London: MIT Press, pp. 291–334.

Birner, R. and Wittmer, H. (2009) 'Making environmental administration more effective: A contribution from new institutional economics', in Beckmann, V. and Palaniswamy, N. (eds), *Institutions and Sustainability*, Heidelberg and Berlin: Springer Verlag, pp. 153–173.

Birner, R., Bhujel, R., Rathgeber, E. M., Sumberg, J., Sriskandarajah, N. and von Sury, F. (2015) *Evaluation of the CGIAR Research Program on Aquatic Agricultural Systems (AAS) Volume 1 – Evaluation Report*, Rome, Italy: Independent Evaluation Arrangement (IEA) of the CGIAR.

Black, J. (1998) 'Regulation as facilitation: Negotiating the genetic revolution', *The Modern Law Review*, 61: 621–660.

BMGF (2013) *Agricultural Development: Strategy Document*, Seattle: Bill & Melinda Gates Foundation. http://www.gatesfoundation.org/agriculturaldevelopment/Documents/agricultural-development-strategy-overview.pdf (accessed 10 October 2012).

Borlaug, N. and Dowswell, C. (1995) 'Mobilising science and technology to get agriculture moving in Africa', *Development Policy Review*, 13(2): 115–129.

Bouis, H. and Islam, Y. (2012) *Delivering Nutrients Widely through Biofortification: Building on Orange Sweet Potato*, Focus 19, Policy Brief 11, Washington, DC: IFPRI.

Boyer, J. and Westgate, M. (2004) 'Grain yields with limited water', *Journal of Experimental Botany*, 55: 2385–2394.

Braun, D. (2003) 'Lasting tensions in research policy-making – a delegation problem', *Science and Public Policy*, 30(5): 309–321.

Brautigam, D. (2009) *The Dragon's Gift: The Real Story of China in Africa*, Oxford: Oxford University Press.

Brooks, S. (2010) *Rice Biofortification: Lessons for Global Science and Development*, London: Earthscan.

Brooks, S. (2013) *Investing in Food Security? Philanthrocapitalism, Biotechnology and Development*, SPRU Working paper, Brighton: University of Sussex.

Brooks, S. and Johnson-Beebout, S. (2012) 'Contestation as continuity? Biofortification research and the CGIAR', in Sumberg, J., Thompson, J. (eds), *Contested Agronomy: Agricultural Research in a Changing World*. London: Routledge.

Brooks, S., Leach, M., Lucas, M. and Millstone, E. (2009) *Silver Bullets, Grand Challenges and the New Philanthropy*, STEPS Working Paper, Brighton: STEPS Centre.

Brossier, J. (1987) 'Système et système de production, note sur les concepts', *Cahier Sciences Humaines (Orstom)*, 23: 377–390.

Burt, C. M. and Styles, S. W. (2007) *Drip and Micro Irrigation Design and Management*. San Luis Obispo, CA, USA: ITRC.

Buttel, F. (1991) 'The restructuring of the America public agricultural research and extension technology transfer system: Implications for agricultural extension', in Rivera, W. M. and Gustafson, D. J. (eds). *Agricultural Extension: Worldwide Institutional Evolution and Forces for Change*, Amsterdam: Elsevier, pp. 43–56.

Cabral, L., Favareto, A., Mukwereza, L. and Amanor, K. (2016) 'Brazil's agricultural politics in Africa: More Food International and the disputed meanings of "family"', *World Development*, 81: 47–60.

Cadisch, G. and Giller, K. E. (eds) (1997) *Driven by Nature: Plant Litter Quality and Decomposition*, Wallingford, UK: CAB International.

Caillé, A. and Vandenberghe, F. (2016) 'Neo-classical sociology: The prospects of social theory today', *European Journal of Social Theory*, 19: 3–20.

Callon, M. (1991) 'Techno-economic networks and irreversibility', in Law, J. (ed.), *A Sociology of Monsters: Essays on Power, Technology and Domination*, London: Routledge, pp. 132–161.

Campos, H., Cooper, M., Edmeades, G., Loffler, C., Schussler, J. and Ibanez, M. (2006) 'Changes in drought tolerance in maize associated with fifty years of breeding for yield in the US corn belt', *Maydica*, 51: 369.

Capra, F. and Mattei, U. (2015) *The Ecology of Law: Toward a Legal System in Tune with Nature and Community*, Oakland: Berrett-Koehler Publishers.

Carney, J. (2008) 'The bitter harvest of Gambian rice policies', *Globalizations*, 5(2): 129–142.

Carney, J. A. and Rosomoff, R. N. (2009) *In the Shadow of Slavery: Africa's Botanical Legacy in the Atlantic World*, Berkeley, CA: University of California Press.

Cartwright, N. and Hardie, J. (2012) *Evidence-based Policy: A Practical Guide to Doing it Better*, Oxford: Oxford University Press.

Castiglioni, P., Warner, D., Bensen, R.J., Anstrom, D.C., Harrison, J., Stoecker, M., Abad, M., Kumar, G., Salvador, S. and D'Ordine, R. (2008) 'Bacterial RNA chaperones confer abiotic stress tolerance in plants and improved grain yield in maize under water-limited conditions', *Plant Physiology*, 147: 446–455.

Castree, N. (2015) 'Geographers and the discourse of an earth transformed: Influencing the intellectual weather or changing the intellectual climate?', *Geographical Research*, 53: 244–254.

Castree, N., Adams, W. M., Barry, J., Brockington, D., Buscher, B., Corbera, E., Demeritt, D., Duffy, R., Felt, U., Neves, K., Newell, P., Pellizzoni, L., Rigby, K., Robbins, P., Robin, L., Rose, D. B., Ross, A., Schlosberg, D., Sorlin, S., West, P., Whitehead, M. and Wynne, B. (2014) 'Changing the intellectual climate', *Nature Climate Change*, 4: 763–768.

Catholic Relief Services, Tanzania (2012) *Marando Bora! Vine beneficiary receives 200 vine cuttings and earns TSH491,000 (~US$327)*. SASHA-Marando Bora.

Cavanagh, C. J. (2016) 'Resilience, class, and the antifragility of capital', *Resilience*: 1–19, doi: 10.1080/21693293.2016.1241474.

Cernea, M. M. (2005) 'Rites of entrance and rights of citizenship: The uphill battle for social research in CGIAR', in Cernea, M. M. and Kassam, A. H. (eds), *Researching the Culture in Agri-Culture: Social Research for International Agricultural Development*, Wallingford: CABI.

CGIAR Consortium Office (2011a) *Agricultural Research in a Changing World. A Strategy and Results Framework for the Reformed CGIAR*. Montpellier: CGIAR Consortium Office.

CGIAR Consortium Office (2011b) *Briefing Paper on Intellectual Property. The Intersection of Public Goods, Intellectual Property Rights, and Partnerships: Maximizing Impact*, Montpellier: CGIAR Consortium Office.

CCIAR Consortium Office (2012) *Financial report 2012*, Montpellier and Washington, DC: CGIAR Consortium Office and CGIAR Fund Office.

CGIAR Consortium Office (2014) *Financial report for the year 2014*, Montpellier and Washington, DC: CGIAR Consortium Office and CGIAR Fund Office.

CGIAR Consortium Office (2015a) *CGIAR Strategy and Results Framework 2016–2030. Redefining How the CGIAR Does Business until 2030*. Montpellier: CGIAR Consortium Office.

CGIAR Consortium Office (2015b) *Draft 2016 CGIAR Financial Plan*. Montpellier: CGIAR Consortium Office.

CGIAR Science Council (2006) *Positioning the CGIAR in the Global Research for Development Continuum*, Rome, Italy: CGIAR Science Council Secretariat.

Chambers, C. and Jiggins, J. (1987a) 'Agricultural research for resource-poor farmers Part I: Transfer-of-technology and farming systems research', *Agricultural Administration and Extension*, 27: 35–52.

Chambers, C. and Jiggins, J. (1987b) 'Agricultural research for resource-poor farmers Part II: A parsimonious paradigm', *Agricultural Administration and Extension*, 27: 109–128.

Chambers, R. (1983) *Rural Development: Putting the Last First*, London: Longman.

Chambers, R. (1994) 'The origins and practice of participatory rural appraisal', *World Development*, 22 (7): 953–969.

Chambers, R. (2008) *Revolutions in Development Inquiry*, London: Earthscan.

Chambers, R., Pacey, A. and Thrupp, L-A. (eds) (1989) *Farmer First: Farmer Innovation and Agricultural Research*. London: Intermediate Technology Publications.

Chapman, R. and Tripp, R. (2003) *Case Studies of Agricultural Extension Programmes Using Privatized Service Provision*, London: AgREN.

Chau, D. C. (2014) *Exploiting Africa: The Influence of Maoist China in Algeria, Ghana, Tanzania*. Annapolis, MD: Naval Institute Press.

Chavez-Tafur, J. (2013) 'SRI is something unprecedented' [Interview with Norman Uphoff], *Farming Matters* [formerly LEISA Magazine], 29(1): 16–19.

Checkland, P. B. (1981) *Systems Thinking, Systems Practice*, Chichester: J. Wiley & Sons.

CIP (2011) *Sweetpotato for Profit and Health Initiative*, http://sweetpotatoknowledge.org/sweetpotato-introduction/overview/sweetpotato-for-profit-and-health-initiative (accessed 20 January 2015).

Cleaver, F. (2002) 'Reinventing institutions: Bricolage and the social embeddedness of natural resource management', *The European Journal of Development Research*, 14: 11–30.

Cleaver, F. (2012) *Development Through Bricolage: Rethinking Institutions for Natural Resource Management*, London: Routledge.

Clements, E. A. and Fernandes, B. M. (2013) 'Land grabbing, agribusiness and the peasantry in Brazil and Mozambique', *Agrarian South: Journal of Political Economy*, 2: 41–69.

Coe, R., Sinclair, F. and Barrios, E. (2014) 'Scaling up agroforestry requires research "in" rather than "for" development', *Current Opinion in Environmental Sustainability*, 6: 73–77.

Collier, P. and Dercon, S. (2014) 'African agriculture in 50 years: Smallholders in a rapidly changing world?', *World Development*, 63: 92–101.

Collinson, M. P. (2000) *A History of Farming Systems Research*, Wallingford: CABI.

Comaroff, J. and Comaroff, J. L. (2012) 'Theory from the South: Or, how Euro-America is evolving toward Africa', *Anthropological Forum*, 22: 113–131.

Conway, G. (1999) *The Doubly Green Revolution: Food for all in the Twenty First Century*, Ithaca: Cornell University Press.

Cooke, B. and Kothari, U. (2001) *Participation: The New Tyranny?*, London: Zed Books.

Coolsaet, B. (2016) 'Towards an agroecology of knowledges: Recognition, cognitive justice and farmers' autonomy in France', *Journal of Rural Studies*, 47, Part A: 165–171.

Coomes, O. T., McGuire, S. J., Garine, E., Caillon, S., Mckey, D., Demeulenaere, E., Jarvis, D., Aistara, G., Barnaud, A. and Clouvel, P. (2015) 'Farmer seed networks make a limited contribution to agriculture? Four common misconceptions', *Food Policy*, 56: 41–50.

Coulthard, G. S. (2014) *Red Skin, White Masks: Rejecting the Colonial Politics of Recognition*, Minneapolis: University of Minnesota Press.

Cullather, N. (2004) 'Miracles of modernization: The green revolution and the apotheosis of technology', *Diplomatic History*, 28(2): 227–254.

Daddieh, C. K. (1994) 'Contract farming in the oil palm industry: A Ghanaian case study', in Little, P. and Watts, M. (eds), *Living Under Contract: Contract Farming and Agrarian Transformation in Sub-Saharan Africa*, Madison: University of Wisconsin Press.

Dalrymple, D. G. (2005) 'Social science knowledge as a public good', in Cernea, M. M. and Kassam, A. H. (eds), *Researching the Culture in Agri-Culture: Social Research for International Agricultural Development*, Wallingford: CABI.

Dalrymple D. G. (2008) 'International agricultural research as a global public good: Concepts, the CGIAR experience, and policy issues', *Journal of International Development*, 20: 347–379.

David, S., McEwan, M., Low, J. (2012) *Improving Women's Access to Quality Sweet Potato Vines*, Nairobi: International Potato Centre.

Davis, M. (2001) *Late Victorian Holocausts: El Niño Famines and the Making of the Third World*, London: Verso Books.

Dawson, N., Martin, A. and Sikor, T. (2016) 'Green revolution in sub-Saharan Africa: Implications of imposed innovation for the wellbeing of rural smallholders', *World Development*, 78: 204–218.

de Jonge, B., Louwaars, N. P. and Kinderlerer, J. (2015) 'A solution to the controversy on plant variety protection in Africa', *Nature Biotechnology*, 33: 487–488.

de Laulanié, H. (1993) 'Le système de riziculture intensive Malgache', *Tropicultura* 11: 1–19.

de Laulanié, H. (2003) *Le Riz à Madagascar: Un développment en dialogue avec les paysans* [Rice in Madagascar: A development in dialogue with the peasants]. Antananarivo, MG: Editions Ambozontany/Editions Karthala.

De Marchi, B. (2003) 'Public participation and risk governance', *Science and Public Policy*, 30: 171–176.

de Monteiro Jales, M., Jank, M. S., Yao, S. and Carter, C. A. (2006) *Agriculture in Brazil and China: Challenges and Opportunities,* Buenos Aires: Institute for the Integration of Latin America and Caribbean.

de Sousa Santos, B. (2007) *Cognitive Justice in a Global World: Prudent Knowledges for a Decent Life,* Lanham, MA: Lexington Books.

de Sousa Santos, B. (2015) *Epistemologies of the South: Justice Against Epistemicide,* Abingdon, UK: Taylor & Francis.

Denison, R. F. and Mcguire, A. M. (2015) 'What should agriculture copy from natural ecosystems?', *Global Food Security,* 4: 30–36.

Devaux, A., Horton, D., Velasco, C., Thiele, G., Lopez, G., Bernet, T., Reinoso, I. and Ordinola, M. (2009) 'Collective action for market chain innovation in the Andes', *Food Policy,* 34(1): 31–38.

DFID (2015) *Promotion of Sustainable Sweet Potato Production and Post-Harvest Management through Farmer Field Schools in East Africa,* London: DFID, available at: https://www.gov.uk/dfid-research-outputs/promotion-of-sustainable-sweet-potato-production-and-post-harvest-management-through-farmer-field-schools-in-east-africa-final-technical-report (accessed 13 January 2016).

Diefenbach, T. (2009) 'New public management in public sector organizations: The dark sides of managerialistic "enlightenment"', *Public Administration,* 87(4): 892–909.

Diwakar, M. C., Kumar, A., Verma, A. and Uphoff, N. (2012) 'Report on the world record SRI yields in Kharif season 2011 in Nalanda District, Bihar State, India', *Agriculture Today,* June: 54–56.

Djansi, E. E. I. (2015) *Lending a Helping Hand? Chinese Economic and Technical Cooperation in Ghana, 1961–1981.* MA thesis, Legon, Ghana: University of Ghana.

Dobermann, A. (2004) 'A critical assessment of the system of rice intensification (SRI)', *Agricultural Systems,* 79(3): 261–281.

Donahue, T. J. and Ochoa Espejo, P. (2016) 'The analytical–Continental divide: Styles of dealing with problems', *European Journal of Political Theory,* 15: 138–154.

Doorenbos, J. and Kassam, A. H. (1979) *Yield Response to Water. FAO Irrigation and Drainage Paper No. 33,* Rome: FAO.

Doorenbos, J. and Pruitt, W. O. (1977) *Crop Water Requirements. FAO Irrigation and Drainage Paper No. 24,* Rome: FAO.

Douthwaite, B., Apgar, J. M., Schwarz, A., McDougall, C., Attwood, S., Senaratna Sellamuttu S. and Clayton, T. (eds) (2015) *Research in Development: Learning from the CGIAR Research Program on Aquatic Agricultural Systems,* Working Paper: AAS-2015-16, Penang, Malaysia: CGIAR Research Program on Aquatic Agricultural Systems.

Douthwaite, B., Keatinge, J. D. H. and Park, J. R. (2001) 'Why promising technologies fail: The neglected role of user innovation during adoption', *Research Policy*, 30(5): 819–836.

Dowd-Uribe, B., Glover, D. and Schnurr, M. A. (2014) 'Seeds and places: The geographies of transgenic crops in the global south', *Geoforum*, 53, 145–148.

Drèze, J. and Sen, A. K. (1989) *Hunger and Public Action*, Oxford: Clarendon Press.

Dunleavy, P. and Hood, C. (1994) 'From old Public Administration to New Public Management', *Public Money & Management*, 14: 9–16.

Dunlop, C. (2000) 'GMOs and regulatory styles', *Environmental Politics*, 9: 149–155.

Dussel, E. (2013) *Ethics of Liberation: In the Age of Globalization and Exclusion*, Durham, NC: Duke University Press.

Duval, A.-M., Gendron, Y. and Roux-Dufort, C. (2015) 'Exhibiting nongovernmental organizations: Reifying the performance discourse through framing power', *Critical Perspectives on Accounting*, 29: 31–53.

Ekboir, J., Douthwaite, B., Sette, C. and Alvarez, S. (2013) *Dealing with Complexity, Adaptability and Continuity in Agricultural Research for Development Organizations, ILAC Working Paper 15*, Rome: Institutional Learning and Change (ILAC) Initiative – c/o Bioversity International.

Ellis, F. (2000) *Rural Livelihoods and Diversity in Developing Countries*, Oxford and New York: Oxford University Press.

El-Namaky, R. and Demont, M. (2013) 'Hybrid rice in Africa: Challenges and prospects', in Wopereis, M., Johnson, D., Ahmadi, N., Tollens, E. and Jalloh, A. (eds), *Realizing Africa's Rice Promise*, Wallingford, UK: CABI Publishing.

European Food Safety Authority (2012) 'Review of the Séralini et al (2012) publication on a 2-year rodent feeding study with glyphosate formulations and GM maize NK603 as published online on 19 September 2012 in Food and Chemical Toxicology', *EFSA Journal*, 10(10): 1–9.

Evenson, R. E. and Gollin, D. (2003) 'Assessing the impact of the Green Revolution, 1960 to 2000', *Science*, 300: 758–762.

Fairhead, J. and Leach, M. (1996) *Misreading the African Landscape: Society and Ecology in a Forest-Savanna Mosaic*, Cambridge: Cambridge University Press.

Fanon, F. (1967) *Black Skin, White Masks*, London: Pluto Press.

FAO (2012) *The State of Insecurity in the World*, Rome: Food and Agriculture Organization of the United Nations.

FAO (2013a) *Review of Food and Agricultural Policies in the United Republic of Tanzania 2005–2011*. Monitoring African Food and Agricultural Policies (MAFAP), Rome: FAO.

FAO (2013b) *The State of Food Insecurity in the World*, Rome: Food and Agriculture Organization of the United Nations.

FAO (2014a) *The State of Food Insecurity in the World*, Rome: Food and Agriculture Organization of the United Nations.

FAO (2014b) *International Year of Family Farming website*, http://www.fao.org/family-farming-2014/en (accessed 22 October 2016) [online].

FAO (2015) *The State of Food Insecurity in the World*, Rome: Food and Agriculture Organization of the United Nations.

FAO (2016) *The State of Food Insecurity in the World*, Rome: Food and Agriculture Organization of the United Nations.

Fingerman, N. N. (2015) *A Study of Brazilian Trilateral Cooperation in Mozambique: The Case of ProSAVANA and ProALIMENTOS*, FAC Working Paper No 113, Brighton: Future Agricultures Consortium.

Folke, C. (2006) 'Resilience: The emergence of a perspective for social–ecological systems analyses', *Global Environmental Change*, 16: 253–267.

Forum for Agricultural Research in Africa (2011) *Transforming African Agriculture through Research, Advisory Services, Education, and Training*. CAADP Pillar IV Strategy and Operational Plan (2011–2013). Accra, Ghana.

Fraser, N. F. (1995) 'From redistribution to recognition? Dilemmas of justice in a "post-socialist" age', *New Left Review*, 212: 68–93.

Fraser, N. and Honneth, A. (2003) *Redistribution Or Recognition?: A Political-philosophical Exchange*, London and New York: Verso.

Freebairn, D. K. (1995) 'Did the green revolution concentrate incomes? A quantitative study of research reports', *World Development*, 23: 265–279.

Frelat, R., Lopez-Ridaura, S., Giller, K. E., Herrero, M., Douxchamps, S., Djurfeldt, A. A., Erenstein, O., Henderson, B., Kassie, M., Paul, B. K., Rigolot, C., Ritzema, R. S., Rodriguez, D., Van Asten, P. J. A. and Van Wijk, M. T. (2016) 'Drivers of household food availability in sub-Saharan Africa based on big data from small farms', *Proceedings of the National Academy of Sciences*, 113: 458–463.

Fresco, L. O. (1984) *Comparing Anglophone and Francophone Approaches to Farming Systems Research and Extension*, Gainsville: Farming Systems Support Project, International Programs, Institute of Food and Agricultural Sciences, University of Florida.

Friis-Hansen, E. (2000) *Agricultural Policy in Africa after Adjustment*, CDR Policy Paper, Copenhagen: Centre for Development Research.

Friis-Hansen, E., Aben, C. and Kidoido, M. (2004) 'Smallholder agricultural technology development in Soroti District: Synergy between NAADS and Farmer Field Schools', *Uganda Journal of Agricultural Sciences*, 9: 250–256.

Fukuyama, F. (2014) *Political Order and Political Decay: From the Industrial Revolution to the Globalization of Democracy*, New York: Farrar, Straus and Girouz.

G8 (Group of Eight Forum Countries) (2009) *L'Aquila Joint Statement on Global Food Security*, http://feedthefuture.gov/sites/default/files/resource/files/afsi_jointstatement_2009.pdf (accessed 5 January 2016).

Gangas, S. (2015) 'From agency to capabilities: Sen and sociological theory', *Current Sociology* (10.1177/0011392115602521)

Gates, R. (2013) 'Innovation from China boosts agricultural development in Africa', *People's Daily Online* 6 May, http://en.people.cn/90780/207187/8233389.html (accessed 10 December 2014).

Gathorne-Hardy, A., Reddy, D. N., Venkatanarayana, M. and Harriss-White, B. (2016) 'System of rice intensification provides environmental and economic gains but at the expense of social sustainability — A multidisciplinary analysis in India', *Agricultural Systems*, 143: 159–168.

Ghana Ministry of Food and Agriculture (2007) *Food and Agricultural Sector Development Policy (FASDEP II)*, Accra, Ghana: MoFA.

Gibson, R., Lyimo, N., Temu, A., Stathers, T., Page, W., Nsemwa, L., Acola, G. and Lamboll, R. (2005) 'Maize seed selection by East African smallholder farmers and resistance to Maize streak virus', *Annals of Applied Biology*, 147: 153–159.

Giddens, A. (1984) *The Constitution of Society: Outline of the Theory of Structuration*, Cambridge: Polity Press.

Giller, K. E., Andersson, J. A., Corbeels, M., Kirkegaard, J., Mortensen, D., Erenstein, O. and Vanlauwe, B. (2015) 'Beyond conservation agriculture', *Frontiers in Plant Science*, 6: Article 870.

Giller, K. E., Tittonell, P., Rufino, M. C., Van Wijk, M. T., Zingore, S., Mapfumo, P., Adjei-Nsiah, S., Herrero, M., Chikowo, R., Corbeels, M., Rowe, E. C., Baijukya, F., Mwijage, A., Smith, J., Yeboah, E., Van Der Burg, W. J., Sanogo, O. M., Misiko, M., De Ridder, N., Karanja, S., Kaizzi, C., K'ungu, J., Mwale, M., Nwaga, D., Pacini, C. and Vanlauwe, B. (2011) 'Communicating complexity: Integrated assessment of trade-offs concerning soil fertility management within African farming systems to support innovation and development', *Agricultural Systems*, 104: 191–203.

Gilligan, D. O., Kumar, N., McNiven, S., Meenakshi, J. V. and Quisumbing, A. (2014) *Bargaining-Power and Biofortification: The Role of Gender in Adoption of Orange Sweet Potato in Uganda*, IFPRI Discussion Paper 01353, Washington, DC: IFPRI.

Glover, D. (2010) 'Is Bt cotton a pro-poor technology? A review and critique of the empirical record', *Journal of Agrarian Change*, 10(4): 482–509.

Glover, D. (2011a) 'A system designed for rice? Materiality and the invention/discovery of the system of rice intensification', *East Asian Science, Technology and Society*, 5(2): 217–237.

Glover, D. (2011b) 'The system of rice intensification: Time for an empirical turn', *NJAS-Wageningen Journal of Life Sciences*, 57(3–4): 217–224.

Glover, D. (2014) 'Of yield gaps and yield ceilings: Making plants grow in particular places', *Geoforum*, 53: 184–194.

Glover, D., Sumberg, J. and Andersson, J. A. (2016) 'The adoption problem; or why we still understand so little about technological change in African agriculture', *Outlook on Agriculture*, 45(1): 3–6.

Golub, P. S. (2013) 'From the New International Economic Order to the G20: How the "global South" is restructuring world capitalism from within', *Third World Quarterly,* 34(6): 1000–1015.

GRAIN, (2015) 'Land and seed laws under attack. Who is pushing changes in Africa?', Barcelona, Spain: AFSA GRAIN.

Green, M. (2012) 'Anticipatory development: Mobilizing civil society in Tanzania', *Critique of Anthropology*, 32: 309–333.

Griffin, K. (1974) *The Political Economy of Agrarian Change: An Essay on the Green Revolution*, Cambridge, MA: Harvard University Press.

Gruening, G. (2001) 'Origin and theoretical basis of new public management', *International Public Management Journal*, 4: 1–25.

Gubbels, P. (1994) 'The role of peasant farmer organization in transforming agricultural research and extension practices in West Africa', in Bagchee, A. (ed.), *Agricultural Extension in Africa*. World Bank Discussion Paper 231. Washington, DC: The World Bank.

Gudynas, E. (2016) 'Beyond varieties of development: Disputes and alternatives', *Third World Quarterly*, 37: 721–732.

Gyasi, E. A. (1992) 'Emergence of a new oil palm belt in Ghana', *Tijdschrift voor Economische en Sociale Geografie*, 83: 39–49.

Haddad, L. (2013) *Ending Undernutrition: Our Legacy to the Post 2015 Generation*, London and Brighton: Children's Investment Fund Foundation and IDS.

Hagenimana, V., Oyunga, A., Low, J., Nojoroge, S. M., Gichuki, S. T. and Kabira, J. (1999) *The Effects of Women Farmers' Adoption and Production of Orange-Flesh Sweet Potatoes: Raising Vitamin A Intake in Kenya*, Research Report Series 3, Washington, DC: ICRW.

Hajer, M. A. (1995) *The Politics of Environmental Discourse: Ecological Modernization and the Policy Process*, Oxford: Oxford University Press.

Hajer, M. (2006) 'Doing discourse analysis: Coalitions, practices, meaning', in den Brink, V. and Metze, T. (eds), *Words Matter in Policy and Planning: Discourse Theory and Methods in the Social Sciences*, Utrecht: Netherlands Graduate School of Urban and Regional Research, pp. 65–74.

Hammond, B., Dudek, R., Lemen, J. and Nemeth, M. (2004) 'Results of a 13 week safety assurance study with rats fed grain from glyphosate tolerant corn', *Food and Chemical Toxicology*, 42: 1003–1014.

Harsh, M. (2005) 'Formal and informal governance of agricultural biotechnology in Kenya: Participation and accountability in controversy surrounding the draft biosafety bill', *Journal of International Development*, 17: 661–677.

Hartmann, A. (2012) 'Scaling up agricultural value chains for pro-poor development', in Linn, J. F. (ed.), *Scaling up in Agriculture, Rural Development and Nutrition, 2020 Vision for Food, Agriculture, and the Environment*, Washington, DC: International Food Policy Research Institute, pp. 16–17.

HarvestPlus (2012) *Disseminating Orange-Fleshed Sweet Potato: Findings from a HarvestPlus Project in Mozambique and Uganda*, Washington, DC: HarvestPlus.

HarvestPlus (2014) *Press Release: Global Policymakers Commit to Scaling-Up Nutritious Foods to Reach Millions*, Washington, DC: HarvestPlus.

Harwood R. R., Place, F., Kassam, A. H. and Gregersen, H. M. (2006) 'International public goods through integrated natural resources management research in CGIAR partnerships', *Experimental Agriculture*, 42: 375–397.

Haug, R., Hella, J. P., Nchimbi-Msolla, S., Mwaseba, D. L. and Synnevag, G. (2016) 'If technology is the answer, what does it take?', *Development in Practice*, 26: 375–386.

Haugen, H. M. (2015) 'Inappropriate processes and unbalanced outcomes: Plant variety protection in Africa goes beyond UPOV 1991 Requirements', *The Journal of World Intellectual Property*, 18: 196–216.

Haverkort, A. J., Boonekamp, P. M., Hutten, R., Jacobsen, E., Lotz, L. A. P., Kessel, G. J. T., Visser, R. G. F. and Van der Vossen, E. A. G. (2008) 'Societal costs of late blight in potato and prospects of durable resistance through cisgenic modification', *Potato Research*, 51(1): 47–57.

Helen Keller International (HKI) (2012) *Orange-Fleshed Sweet Potato Situation Analysis and Needs Assessment Tanzania Report*, Dar es Salaam: HKI.

Hellin, J. and Meijer, M. (2006) 'Guidelines for value chain analysis', http://www.fao.org/fileadmin/templates/esa/LISFAME/Documents/Ecuador/value_chain_methodology_EN.pdf (accessed 14 October 2013).

Hengsdijk, H., Franke, A. C., Van Wijk, M. T. and Giller, K. E. (2014) *How Small is Beautiful? Food Self-Sufficiency and Land Gap Analysis of Smallholders in Humid and Semi-Arid Sub Saharan Africa*, Report 562, Plant Research International, Wageningen: Wageningen University.

Henke, C. R. (2000) 'Making a place for science: The field trial', *Social Studies of Science*, 30(4): 483–511.

Herring, R. J. (2007) 'The genomics revolution and development studies: Science, poverty and politics', *Journal of Development Studies*, 43: 1–30.

Hirvonen, O. and Laitinen, A. (2016) 'Recognition and democracy – An introduction', *Thesis Eleven*, 134: 3–12.

HLPE (2011) *Price Volatility and Food Security Report by the High Level Panel of Experts on Food Security and Nutrition of the Committee on World Food Security*, Rome: UN Food and Agriculture Organization.

Hoffman, H. (1982) 'Towards Africa? Brazil and the South-South trade', in Carlsson, J. (ed.), *South-South Relations in a Changing World*, Uppsala: Scandinavian Institute of African Studies, pp. 55–77.

Holston, J. (2008) *Insurgent Citizenship: Disjunctions of Democracy and Modernity in Brazil*, Princeton: Princeton University Press.

Honneth, A. (1996) *The Struggle for Recognition: The Moral Grammar of Social Conflicts*, Cambridge MA: MIT Press.

Honneth, A. (2014) *Freedom's Right: The Social Foundations of Democratic Life*, New York: Columbia University Press.

Hotz, C., Loechl, C., Lubowa, A., Tumwine, J., Ndeezi, G., Nandutu Masawi, A., Baingana, R., Carriquiry, A., de Brauw, A., Meenakshi, J. V. and Gilligan, D. (2012) 'A large scale intervention to introduce beta carotene rich orange sweet potato was effective in increasing vitamin A intakes among children and women in rural Uganda', *Journal of Nutrition*, 142: 1871–1880.

Hounkonnou, D., Kuyper, T. W., Kossou, D., Leeuwis, C., Nederlof, S., Röling, N., Sakyi-Dawson, O., Traoré, M. and van Huis, A. (2012) 'An innovation systems approach to institutional change: Smallholder development in West Africa', *Agricultural Systems*, 108: 74–83.

Howell, K. R., Shrestha, P. and Dodd, I. C. (2015) 'Alternate wetting and drying irrigation maintained rice yields despite half the irrigation volume, but is currently unlikely to be adopted by smallholder lowland rice farmers in Nepal', *Food and Energy Security*, 4: 144–157.

Hulme, M. (2014) *Can Science Fix Climate Change: A Case Against Climate Engineering*, Cambridge: Polity Press.

Humphries, S., Rosas, J. C., Gómez, M., Jiménez, J., Sierra, F., Gallardo, O., Avila, C. and Barahona, M. (2015) 'Synergies at the interface of farmer–scientist partnerships: Agricultural innovation through participatory research and plant breeding in Honduras', *Agriculture & Food Security*, 4(1): 1–17.

Hunsberger, C. (2010) 'Jatropha as a biofuel crop and the economy of appearances: Experiences from Kenya', *Review of African Political Economy*, 41(140): 216–231.

Hutchby, I. (2001) 'Technologies, texts and affordances', *Sociology*, 35(2): 441–456.

Hutchins, E. (1996) 'Learning to navigate', in Chaiklin, S. and Lave, J. (eds), *Understanding Practice; Perspectives on Activity and Context*, Cambridge: Cambridge University Press, pp. 35–63.

IAASTD (2009) *International Assessment of Agricultural Knowledge, Science and Technology for Development: Global Report*, eds McIntyre, B. D., Herren, H. R., Wakhungu, J. and Watson, R. T., Washington, DC: Island Press.

IFPRI (2004) *Impact Assessment of the National Agricultural Advisory Services*, Washington, DC: IFPRI.

Igoe, J. (2010) 'The spectacle of nature in the global economy of appearances: Anthropological engagements with the spectacular mediations of transnational conservation', *Critique of Anthropology*, 30(4): 375–397.

Ingram, J., Ericksen, P. and Liverman, D. (eds) (2010) *Food Security and Global Environmental Change*, London: Earthscan.

International Food Policy Research Institute (IFPRI) (2014) *2014 Nutrition Country Profile: United Republic of Tanzania*, Washington, DC: IFPRI.

ISSC, IDS and UNESCO (2016), *World Social Science Report 2016, Challenging Inequalities: Pathways to a Just World*, UNESCO Publishing: Paris.

ITAD Ltd (2008) *Performance Evaluation of National Agricultural Advisory Services*, Final Report, Hove: ITAD.

Jackson, M. C. (1985) 'Social systems theory and practice: The need for a critical approach', *International Journal of General Systems*, 10: 135–151.

Jank, M. S., Franco, M., Leme, P., Nassar, A. M. and Filho, P. F. (2001) 'Concentration and internationalization of Brazilian agribusiness exporters', *International Food and Agribusiness Management Review*, 2(3/4): 359–374.

Jansen, K. and Gupta, A. (2009) 'Anticipating the future: "Biotechnology for the poor" as unrealized promise?', *Futures*, 41(7): 436–445.

Jarvis, D. I. and Hodgkin, T. (1999) 'Wild relatives and crop cultivars: Detecting natural introgression and farmer selection of new genetic combinations in agroecosystems', *Molecular Ecology*, 8: S159–S173.

Jasanoff, S. (1995) 'Product, process, or programme: Three cultures and the regulation of biotechnology', in Bauer, M. (ed.), *Resistance to New Technology: Nuclear Power, Information Technology and Biotechnology*, Cambridge: Cambridge University Press.

Jasanoff, S. (2003) 'Technologies of humility: Citizen participation in governing science', *Minerva*, 41: 223–244.

Jones, G. E. and Garforth, C. (1997) 'Chapter 1 - The history, development, and future of agricultural extension', in Swanson, B. E., Bentz, R. P. and Sofranko, A. J. (eds), *Improving Agricultural Extension. A Reference Manual*, Rome: Food and Agricultural Organization.

Jones, K. (2013) 'The political ecology of market-oriented seed system development and emergent alternatives', in *Food Sovereignty: A Critical Dialogue, International Conference*, Yale University, 14 September 2013.

Jones, K. M. and de Brauw, A. (2015) 'Using agriculture to improve child health: Promoting orange sweet potatoes reduces diarrhea', *World Development*, 74: 15–24.

Jones, R. B., Bramel, P., Longley, C. and Remington, T. (2002) 'The need to look beyond the production and provision of relief seed: Experiences from southern Sudan', *Disasters*, 26: 302–315.

Joughin, J. and Kjaer, A. M. (2010) 'The politics of agricultural policy reforms: The case of Uganda', *Forum for Development Studies*, 37(1): 61–78.

Jouve, P. (1986) 'Quelques principes de construction de typologies d'exploitations agricoles suivant différentes situations agraires', *Les cahiers de la Recherche-Développement*, 11: 48–56.

Kamanda, J. O. (2015) *Activating Institutional Innovations for Hunger and Poverty Reduction: Potential of Applied International Agricultural Research*, PhD thesis, Faculty of Agricultural Sciences, University of Hohenheim.

Kapinga, R., Andrade, M., Tumwegamire, S., Lemaga, B., Agili, S., Marandu, E., Laurie, S., Ndolo, P. and Mwanga, R. (2006) *Evaluation and Large-Scale Dissemination of Orange-Fleshed Sweetpotato in sub Saharan Africa*, Powerpoint presentation, ISTRC Fourteenth Triennial Symposium - Thiruvanathapuram, Kerala, India.

Kapinga, R., Ewell, P., Jermiah, C. Simon and Kileo, R. (1995) *Sweet Potato in Tanzanian Farming and Food Systems: Implications for Research*, Dar Es Salaam: International Potato Centre, Ministry of Agriculture, Tanzania.

Karembu, M., Otunge, D. and Wafula, D. (2010) *Developing a Biosafety Law: Lessons from the Kenyan Experience*, Nairobi: ISAAA AfriCentre.

Kassam, A. H. (2006) 'Receptivity to social research by the CGIAR', *Experimental Agriculture*, 42(2): 217–228.

Kassam, A., Stoop, W. and Uphoff, N. (2011) 'Review of SRI modifications in rice crop and water management and research issues for making further improvements in agricultural and water productivity', *Paddy and Water Environment*, 9(1): 163–180.

Kassam, A. H., Gregersen, H. M., Fereres, E., Javier, E. Q., Harwood, R. R., De Janvry, A. and Cernea, M. M. (2004) 'A framework for enhancing and guarding the relevance and quality of science: The case of the CGIAR', *Experimental Agriculture*, 40: 1–21.

Keeley, J. and Scoones, I. (2003) *Understanding Environmental Policy Processes: Cases From Africa*, London: Earthscan.

Kerr, R. B. (2012) 'Lessons from the old Green Revolution for the new: Social, environmental and nutritional issues for agricultural change in Africa', *Progress in Development Studies*, 12: 213–229.

Kimenju, S. C. and De Groote, H. (2008) 'Consumer willingness to pay for genetically modified food in Kenya', *Agricultural Economics*, 38: 35–46.

Kimura, A. H. (2013) *Hidden Hunger: Gender and the Politics of Smarter Foods*, New York: Cornell University Press.

Kingiri, A. N. (2010) 'Experts to the rescue? An analysis of the role of experts in biotechnology regulation in Kenya', *Journal of International Development*, 22: 325–340.

Kjaer, A. M. and Joughin, J. (2012) 'The reversal of agricultural extension reform in Uganda: Ownership and values', *Policy and Society*, 31(4): 319–330.

Klerkx, L. and Leeuwis, C. (2008) 'Delegation of authority in research funding to networks: Experiences with a multiple goal boundary organization', *Science and Public Policy*, 35(3): 183–196.

Klerkx, L. W. A., Hall, A. and Leeuwis, C. (2009) 'Strengthening agricultural innovation capacity: Are innovation brokers the answer?', *International Journal of Agricultural Resources, Governance and Ecology*, 8(5–6): 409–438.

Kloppenburg, J. (2010) 'Impeding dispossession, enabling repossession: Biological open source and the recovery of seed sovereignty', *Journal of Agrarian Change*, 10: 367–388.

Kloppenburg, J. (2014) 'Re-purposing the master's tools: The open source seed initiative and the struggle for seed sovereignty', *Journal of Peasant Studies*, 41: 1225–1246.

Konings, P. (1986) *The State and Rural Class Formation in Ghana: A Comparative Analysis*, London: Routledge and Kegan Paul.

Kragelund, P. (2015) 'Towards convergence and cooperation in the global development finance regime: Closing Africa's policy space?', *Cambridge Review of International Affairs*, 28(2): 246–262.

Kranjac-Berisavljevic, G., Blench, R. M. and Chapman, R. (2003) *Multi-Agency Partnerships (MAPs) for Technical Change in West African Agriculture: Rice Production and Livelihoods in Ghana*, London, UK: Overseas Development Institute.

Kristjanson, P., Reid, R. S., Dickson, N., Clark, W. C., Romney, D., Puskur, R., MacMillan, S. and Grace, D. (2009) 'Linking international agricultural research knowledge with action for sustainable development', *Proceedings of the National Academy of Sciences*, 9(13): 5047–5052.

Kurimoto, E. (1984) 'Agriculture in the multiple subsistence economy of the Pari', in Sakamoto, K. (ed.), *Agriculture and Land Utilization in the Eastern Zaire and the Southern Sudan*, Kyoto, Japan: Department of Agriculture and Forestry Economics, Faculty of Agriculture, Kyoto University, pp. 23–52.

Kurimoto, E. (1995) Coping with enemies: Graded age system among the Pari of Southeastern Sudan. Osaka, Japan: Osaka University, pp. 261–311.

Kuyper, T. W. and Struik, P. C. (2014) 'Epilogue: Global food security, rhetoric, and the sustainable intensification debate', *Current Opinion in Environmental Sustainability*, 8: 71–79.

Langyintuo, A. S., Mwangi, W., Diallo, A. O., MacRobert, J., Dixon, J. and Banziger, M. (2010) 'Challenges of the maize seed industry in eastern and southern Africa: A compelling case for private-public intervention to promote growth', *Food Policy*, 35: 323–331.

Latour, B. (1991) 'Technology is society made durable', in Law, J. (ed.), *A Sociology of Monsters: Essays on Power, Technology and Domination*, London: Routledge, pp. 103–131.

Latour, B. (1992) 'Where are the missing masses? The sociology of a few mundane artifacts', in Bijker, W. E and Law, J. (eds), *Shaping Technology/ Building Society: Studies in Sociotechnical Change.* Cambridge, MA, MIT Press, pp. 225–258.

Leach, M. and Mearns, R. (1996) *The Lie of the Land: Challenging Received Wisdom on the African Environment,* London: The International Africa Institute.

Leach, M., Scoones, I. and Stirling, I. (2010) *Dynamic Sustainabilities: Technology, Environment, Social Justice,* London: Earthscan.

Leeuwis, C. (2013) 'Coupled performance and change in the making', Inaugural lecture, Wageningen University, Wageningen.

Leeuwis, C. (with contributions by A. Van den Ban) (2004) *Communication for Rural Innovation. Rethinking Agricultural Extension,* Oxford: Blackwell Science.

Leeuwis, C. and S. Wigboldus (2017) 'What kinds of "systems" are we dealing with? Implications for systems research and scaling', in Öborn, I., Vanlauwe, B., Phillips, M., Thomas, R., Brooijmans, W. and Atta-Krah, K. (eds), *Sustainable Intensification in Smallholder Agriculture. An Integrated Systems Research Approach*, London: Routledge Taylor & Francis Group.

Leeuwis, C., Schut, M., Waters-Bayer, A., Mur, R., Atta-Krah, K. and Douthwaite, B. (2014) *Capacity to innovate from a system CGIAR research program perspective,* Program Brief: AAS-2014-29, Penang, Malaysia: CGIAR Research Program on Aquatic Agricultural Systems.

Leftwich, A. (1995) 'Bringing politics back in: Towards a model of the developmental state', *The Journal of Development Studies,* 31: 400–427.

Levidow, L., Carr, S. and Wield, D. (2000) 'Genetically modified crops in the European Union: Regulatory conflicts as precautionary opportunities', *Journal of Risk Research,* 3: 189–208.

Lévi-Strauss, C. (1955) 'The structural study of myth', *The Journal of American Folklore,* 68: 428–444.

Levi-Strauss, C. (1966) *The Savage Mind,* Chicago: University of Chicago Press.

Li, J., Xin, Y. and Yuan, L. (2009) *Hybrid Rice Technology: Ensuring China's Food Security.* IFPRI Discussion Paper 00918, Washington, DC: International Food Policy Research Institute.

Li, T. (2011) 'Rendering society technical: Government through community and the ethnographic turn at the World Bank in Indonesia', in Mosse, D. (ed.), *Adventures in Aidland: The Anthropology of Professionals in International Development,* Oxford: Berghahn, pp. 57–80.

Loevinsohn, M. and Kaiser, K. (1982) *Recommendations and Farmer's Practices. IRRI Working Paper.* Los Banos, Philippines: International Rice Research Institute.

Löfstedt, R. E. and Vogel, D. (2001) 'The changing character of regulation: A comparison of Europe and the United States', *Risk Analysis,* 21: 399–416.

Long, N. (2001) *Development Sociology: Actor Perspectives,* London: Routledge.

Long, N. and Long, A. (eds) (1992) *Battlefields of Knowledge: The Interlocking of Theory and Practice in Social Research and Development*, London: Routledge.

Loos, J., Abson, D. J., Chappell, M. J., Hanspach, J., Mikulcak, F., Tichit, M. and Fischer, J. (2014) 'Putting meaning back into "sustainable intensification"', *Frontiers in Ecology and the Environment*, 12: 356–361.

Louette, D., Charrier, A. and Berthaud, J. (1997) 'In situ conservation of maize in Mexico: Genetic diversity and maize seed management in a traditional community', *Economic Botany*, 51: 20–38.

Louwaars, N. P. and de Boef, W. S. (2012) 'Integrated seed sector development in Africa: A conceptual framework for creating coherence between practices, programs, and policies', *Journal of Crop Improvement*, 26: 39–59.

Lövbrand, E., Beck, S., Chilvers, J., Forsyth, T., Hedrén, J., Hulme, M., Lidskog, R. and Vasileiadou, E. (2015) 'Who speaks for the future of Earth? How critical social science can extend the conversation on the Anthropocene', *Global Environmental Change*, 32: 211–218.

Low, J., Arimond, M., Labarta, R., Andrade, M. and Namanda, S. (2013) 'The introduction of orange-fleshed sweet potato in Mozambican diets: A marginal change to make a major difference', in Fanzo, J., Hunter, D., Borelli, T. and Mattei, F. (eds), *Diversifying Food and Diets: Using Agricultural Biodiversity to Improve Nutrition and Health*, Earthscan: New York, pp. 283–290.

Lowder, S. K., Skoet, J. and Singh, S. (2014) *What Do We Really Know About the Number and Distribution of Farms and Family Farms in The World? Background Paper for the State of Food and Agriculture*, ESA Working Paper 14-02, Rome: FAO.

Luhmann, N. (1984) *Soziale Systeme: Grundriss einer allgemeinen Theorie,* Suhrkamp Taschenbuch Wissenschaft 666, Frankfurt am Main: Suhrkamp.

Ly, P., Jensen, L. S., Bruun, T. B., Rutz, D. and de Neergaard, A. (2012) 'The system of rice intensification: Adapted practices, reported outcomes and their relevance in Cambodia', *Agricultural Systems*, 113: 16–27.

MAAIF (Ministry of Agriculture, Animal Industry and Fisheries) (2000) The National Agricultural Advisory Services (NAADS) Programme - Master Document of the NAADS Task Force and Joint Donor Group, Entebbe: MAAIF.

MAAIF (Ministry of Agriculture, Animal Industry and Fisheries) (2002) A Functional Analysis Report of the Ministry of Agriculture, Animal Industry and Fisheries, Entebbe, MAAIF.

MAAIF (Ministry of Agriculture Animal Industry and Fisheries) (2009) The Policy Position of the Ministry of Agriculture Animal Industry and Fisheries on the Proposed Conversion of Public Extension Staff in Local Governments to National Agricultural Advisory Services. November 2009, Entebbe, Uganda.

Maat, H. (2015) 'Commodities and anti-commodities: Rice on Sumatra 1915-1925', in Edda Fields Black, Francesca Bray, Dagmar Schäfer and Peter Coclanis (eds), *Rice; Global Networks, New Histories*, New York: Cambridge University Press, pp. 335–354.

Maat, H. and Glover, D. (2012) 'Alternative configurations of agricultural experimentation', in Sumberg, J. and Thompson, J. (eds), *Contested Agronomy: Agricultural Research in a Changing World*, London: Routledge, pp. 131–145.

Maat, H. and Hazareesingh, S. (eds) (2015) *Local Subversions of Colonial Cultures Commodities and Anti-Commodities in Global History*, Basingstoke & New York: Palgrave Macmillan.

Mangheni, N. M., Mutimba, J. and Biryabaho, F. M. (2003) *Responding to the Shift from Public to Private Contractual Agricultural Extension Service Delivery: Educational Implications of*

Policy Reforms in Uganda, Paper presented at the 19th Annual AIAEE Conference, Raleigh, North Carolina, 8–12 April.

Mann, G. (2015) *From Empires to NGOs in the West African Sahel: The Road to Nongovernmentality*, New York: Cambridge University Press.

Manning, N. (2001) 'The legacy of the New Public Management in developing countries', *International Review of Administrative Sciences*, 67: 297–312.

Martin, A., Coolsaet, B., Corbera, E., Dawson, N. M., Fraser, J. A., Lehmann, I. and Rodriguez, I. (2016) 'Justice and conservation: The need to incorporate recognition', *Biological Conservation*, 197: 254–261.

MASA (2015) 'Master plan for agrarian development along the Nacala Corridor, Mozambique'. Ministry of Agriculture and Food Security, Government of Mozambique, available at: http://www.prosavana.gov.mz/lib-master-plan/ (accessed 15 February 2017).

Maturana, H. R. and Varela, F. J. (1984) *The Tree of Knowledge: The Biological Roots of Human Understanding*, Boston: Shambala.

Mauser, W., Klepper, G., Rice, M., Schmalzbauer, B. S., Hackmann, H., Leemans, R. and Moore, H. (2013) 'Transdisciplinary global change research: The co-creation of knowledge for sustainability', *Current Opinion in Environmental Sustainability*, 5(3–4): 420–431.

Mawdsley, E. (2015) 'Development Geography 1: Cooperation, competition and convergence between "North" and "South"', *Progress in Human Geography*, Sage Journal Online, 0309132515601776, first published on 27 August, as doi:10.1177/ 0309132515601776.

Mazoyer, M. *et al.* (2002) *Larousse Agricole*, 4th edn, Paris: Larousse.

Mbabu, A., Munyua, H., Mulongo, G., David, S. and Brendin, M. (2015) *Learning the Smart Way: Lessons Learned by the Reaching Agents of Change Project*, Nairobi, Kenya: International Potato Center.

McCalla, A. F. (2014) *CGIAR Reform – Why So Difficult? Review, Reform, Renewal, Restructuring, Reform Again and then 'The New CGIAR' – So Much Talk and So Little Basic Structural Change - Why?*, University of California, Davis: Department of Agricultural and Resource Economics.

McDermott, J., Aït-Aïssa, M., Morel J. and Rapando, N. (2013) 'Agriculture and household nutrition security – development practice and research needs', *Food Security*, 5: 667–678.

McDonald, A. J., Hobbs, P. R. and Riha, S. J. (2006) 'Does the system of rice intensification outperform conventional best management? A synopsis of the empirical record', *Field Crops Research*, 96(1): 31–36.

McDonald, A. J., Hobbs, P. R. and Riha, S. J. (2008) 'Stubborn facts: Still no evidence that the System of Rice Intensification out-yields best management practices (BMPs) beyond Madagascar', *Field Crops Research*, 108(2): 188–191.

McEwan, C. and Mawdsley, E. (2012) 'Trilateral development cooperation: Power and politics in emerging aid relationships', *Development and Change*, 43(6): 1185–1209.

McFeat, T. (1972) *Small-Group Cultures*, New York: Pergamon Press.

McGuire, S. and Sperling, L. (2013) 'Making seed systems more resilient to stress', *Global Environmental Change: Human and Policy Dimensions*, 23: 644–653.

McGuire, S. and Sperling, L. (2016) 'Seed systems smallholder farmers use', *Food Security*, 8(1): 179–195.

McMichael, P. (1997) 'Rethinking globalization: The agrarian question revisited', *Review of International Political Economy*, 4(4): 630–62.

McNie, E. C., Parris, A. and Sarewitz, D. (2016) 'Improving the public value of science: A typology to inform discussion, design and implementation of research', *Research Policy*, 45(4): 884–895.

Meek, D. and Simonian, L. L. (2016) 'Transforming space and society? The political ecology of education in the Brazilian Landless Workers' Movement's Jornada de Agroecologia', *Environment and Planning D: Society and Space* (doi:10.1177/0263775816667073).

Meinzen-Dick, R., Quisumbing, A., Behrman, J., Biermayr-Jenzano, P., Wilde, V., Noordeloos, M., Ragasa, C. and Beintema, M. (2011) *Engendering Agricultural Research*, IFPRI Research Monograph, Washington, DC: International Food Policy Research Institute.

Mekbib, F. (2006) 'Farmer and formal breeding of sorghum (Sorghum bicolor (L.) Moench) and the implications for integrated plant breeding', *Euphytica*, 152: 163–176.

Merrey, D. J., McLeod, R. and Szonyi, J. (2015) *Evaluation of the CGIAR Research Program on 'Dryland Systems'*, Amman, Jordan: ICARDA.

Mintz, S. W. (1986) *Sweetness and Power: The Place of Sugar in Modern History*, London: Penguin Books.

Mkindi, A. R. (2015) *Farmers' Seed Sovereignty Is Under Threat*, Policy paper 03/2015, Berlin: Rosa-Luxemburg-Stiftung.

Mmasa, J., Mbaula, W., Swai, S., Maro, F. E. and Mpenda, Z. (2014) *Sweet Potato Value Chain Actors Mapping Study: A Case of Magu District, Doroma*, Tanzania: University of Dodoma.

Moore, J. W. (2015) *Capitalism in the Web of Life: Ecology and the Accumulation of Capital*, London: Verso Books.

Moseley, W. G. (1993) *Indigenous Agroecological Knowledge Among the Bambara of Djitoumou, Mali: Foundation for a Sustainable Community*. M.Sc. Thesis, School of Natural Resources, University of Michigan, Ann Arbor, MI USA.

Moseley, W. G. (1995) 'Securing livelihoods in marginal environments: Can NGOs make a long term difference?', in *Policy In The Making, Poverty and Food Economy: Assessing Livelihoods*, Save the Children Fund, London. Policy Development Unit Discussion Paper Number 4, pp. 12–22.

Moseley, W. G. (1996) 'A foundation for coping with environmental change: Indigenous agroecological knowledge among the Bambara of Djitoumou, Mali', in Adams, W. M. and Slikkerveer, J. (eds), *Indigenous Knowledge and Change in African Agriculture*, Studies in Technology and Social Change Series, No. 26, Ames, Iowa: Iowa State University, pp. 11–130.

Moseley, W. G. (2007) 'Collaborating in the field, working for change: Reflecting on partnerships between academics, development organizations and rural communities in Africa', *Singapore Journal of Tropical Geography*, 28: 334–347.

Moseley, W. G. (2008) 'Mali's cotton conundrum: Commodity production and development on the periphery', in Moseley, W. G. and Gray, L. C. (eds), *Hanging by a Thread: Cotton, Globalization and Poverty in Africa,* Athens, OH: Ohio University Press, pp. 83–102.

Moseley, W. G. and Gray, L. C. (eds) (2008) *Hanging by a Thread: Cotton, Globalization and Poverty in Africa*, Athens, OH: Ohio University Press and Nordic Africa Press.

Moseley, W. G., Bangaly, M., Diallo, M. and Coulibaly, M. (1994) *Rapport d'Evaluation sur La Banque Grameen de L'Association des Femmes d'Attara*, Mimeo. Mopti, Mali: Save the Children Fund (UK).

Moseley, W. G., Perramond, E., Hapke H. and Laris, P. (2013) *An Introduction to Human-Environment Geography: Local Dynamics and Global Processes*, Hoboken, NJ: Wiley/Blackwell.

Moseley, W. G., Schnurr, M. and Bezner Kerr, R. (2015) 'Interrogating the technocratic (neoliberal) agenda for agricultural development and hunger alleviation in Africa', *African Geographical Review*, 34(1): 1–7.

Moshi, A. J. and Marandu, W. (1985) 'Maize research in Tanzania, to feed ourselves', Proceedings of the first eastern, central and southern Africa Regional Maize Workshop, Lusaka, Zambia, 10-17 March, 1985, Lusaka, Zambia: CAB Direct.

Mosse, D. (2005) *Cultivating Development: An Ethnography of Aid Policy and Practice*, London and Ann Arbor, MI: Pluto Press.

Murdoch, J. (1997) 'Towards a geography of heterogeneous associations', *Progress in Human Geography*, 21(3): 321–337.

Murrell, P. (1992) 'Evolutionary and radical approaches to economic reform', *Economics of Planning*, 25(1): 79–95.

Musemakweri, J. (2007) *Farmers' Experiences and Perceptions of the NAADS Agricultural Extension System/Program in Kabale District, Uganda*, Thesis. Iowa State University, USA.

Nally, D. P. (2011a) 'The biopolitics of food provisioning', *Transactions of the Institute of British Geographers*, 36: 37–53.

Nally, D. P. (2011b) *Human Encumbrances: Political Violence and the Great Irish Famine*, Notre Dame, IN: University of Notre Dame Press.

Namanda, S., Potts, M. J., Agili, S., Mwenda, B., Ekinyu, E. and Mwanga, R. (2006) 'Commercialization of sweet potato planting material production to address the "Hunger Gap" in East Africa', Powerpoint presentation. ISTRC Fourteenth Triennial Symposium - Thiruvanathapuram, Kerala, India.

National Agricultural Advisory Services (NAADS) Act (2001) Acts Supplement No. 9 to The Uganda Gazette No. 33 Volume XCIV dated 1 June 2001. Printed by UPPC, Entebbe.

Negin, J., Remans, R., Karuti, S. and Fanzo, J. (2009) 'Integrating a broader notion of food security and gender empowerment into the African Green Revolution', *Food Security*, 1: 351–360.

Neuchâtel Group (1999) *Common Framework on Agricultural Extension*, Paris: Neuchâtel Group.

Neuchâtel Group (2002) *Common Framework on Financing Agricultural and Rural Extension*, Paris: Neuchâtel Group.

Newell, P. (2002) *Biotechnology and the Politics of Regulation*, IDS Working Paper 146. Brighton: IDS.

Ngwediagi, P., Maeda, E., Kimomwe, H., Kamara, R., Massawe, S., Akonaay, H. B. and Mapunda, L. N. D. (2009) *Tanzania Country Report on the State of Plant Genetic Resources for Food and Agriculture*. Rome: FAO.

Noltze, M., Schwarze, S. and Qaim, M. (2012) 'Understanding the adoption of system technologies in smallholder agriculture: The System of Rice Intensification (SRI) in Timor Leste', *Agricultural Systems*, 108: 64–73.

North, D. C. (1990) *Institutions, Institutional Change and Economic Performance*, Cambridge: Cambridge University Press.

Nunan, F. (2015) *Understanding Poverty and the Environment: Analytical Frameworks and Approaches*, London: Routledge.

Nussbaum, M. C. (2009) *Frontiers of Justice: Disability, Nationality, Species Membership*, Cambridge MA: Harvard University Press.

Nussbaum, M. C. (2011) *Creating Capabilities*, Cambridge MA: Harvard University Press.

Obaa, B., Mutimba, J. and Semana, A. R. (2005) *Prioritizing Farmers' Extension Needs in a Publicly-funded Contract System: A Case Study from Mukono District, Uganda*, AgREN, paper No. 147, London: AgREN.

Obirih-Opareh, N. (2008) 'Socio-economic analysis of rice production in Ghana: Agenda for policy study', *Ghana Journal of Agricultural Science*, 41(2): 203–212.

OECD (1998) *Test No. 408: Repeated Dose 90-Day Oral Toxicity Study in Rodents, OECD Guidelines for the Testing of Chemicals Section 4*, Paris: OECD Publishing.

OFAB (2007) Proceedings of stakeholders half-day workshop on the Biosafety Bill 2007, Held in Jacaranda Hotel, Nairobi.

Ogero, K., McEwan, M., Lusheshanija, D. and Mayanja O. (2015) *Can Farmer Multipliers Meet QDS Standards in Production of Sweetpotato Planting Material?*, Lima, Peru: International Potato Centre

Ogunsanwo, A. (1974) *China's Policy in Africa, 1958–71*, Cambridge, UK: Cambridge University Press.

Okusu, H. (2009) 'Biotechnology research in the CGIAR: an overview', *AgBioForum*, 12: 70–77.

Orr, A. (2012) 'Why were so many social scientists wrong about the Green Revolution? Learning from Bangladesh', *Journal of Development Studies*, 48: 1565–1586.

Osei, R. D (2012) *GADCO: A Holistic Approach to Tackling Low Agricultural Incomes: Growing Inclusive Markets*, UNDP.

Ostrom, E. (2015) *Governing the Commons*, Cambridge: Cambridge University Press.

Ostrom, E., Gardner, R. and Walker, J. (1994) *Rules, Games, and Common-Pool Resources*, Ann Arbor: The University of Michigan Press.

Otsuka, K., Cordova, V. and David, C. C. (1992) 'Green revolution, land reform, and household income distribution in the Philippines', *Economic Development and Cultural Change*, 40: 719–741.

Owino, O. (2012) 'Scientists torn over Kenya's recent GM food ban', SciDev.Net 30 November 2011. Available at: http://www.scidev.net/global/nutrition/news/scientists-torn-over-kenya-s-recent-gm-food-ban.html (accessed 10 February 2017).

Oxford English Dictionary (2012) 'agronomy, n.', Oxford University Press.

Paarlberg, R. L. (2001) *The Politics of Precaution: Genetically Modified Crops in Developing Countries*, Washington, DC: IFPRI.

Page, G. (2012) 'How to ensure the world's food supply', *The Washington Post*. 2 August.

Parkinson, S. (2009) 'When farmers don't want ownership: Reflection on demand-driven extension in Sub-Saharan Africa', *Journal of Agricultural Education and Extension*, 15(4): 417–429.

Parsons, T. (1951) *The Social System*, Glencoe: Free Press.

Patel, R. (2009) *Stuffed & Starved: Markets, Power & the Hidden Battle for the World Food System*, London: Portobello.

Patel, R. (2013) 'The long green revolution', *The Journal of Peasant Studies*, 40: 1–63.

Patriota, T. C. and Pierri, F. M. (2014) 'Brazil's cooperation in African agricultural development and food security', in Cheru, F. and Modi, R. (eds), *Agricultural Development and Food Security in Africa: The Impact of Chinese, Indian and Brazilian Investment*, London: Zed Books, pp. 125–144.

Peking Review (1962) 'Sino-Ghanaian co-operation protocol', *Peking Review*, 43: 23.

Perry, C. (2007) 'Efficient irrigation, inefficient communication, flawed recommendations', *Irrigation and Drainage*, 56: 367–378.

Pfaffenberger, B. (1992) 'Social anthropology of technology', *Annual Review of Anthropology*, 21(1): 491–516.

Pingali, P. (2010) 'Global agriculture R&D and the changing aid architecture', *Agricultural Economics*, 41(Supp. 1): 145–153.

Pingali, P. L. (2012) 'Green revolution: Impacts, limits, and the path ahead', *Proceedings of the National Academy of Sciences*, 109: 12302–12308.

Pinstrup-Andersen, P. (2005) 'Note from Chairman of the Science Council', CGIAR Newsletter, CGIAR (October 2005).

Pippin, R. B. (1991) *Modernism as a Philosophical Problem: On the Dissatisfactions of European High Culture*, Oxford: Basil Blackwell.

Pittore, K. and Robinson, E. (2015) 'Food, markets and nutrition: Maximizing the impacts of private sector engagement in Tanzania: Case studies and key messages from the workshop', Brighton: IDS.

PMA (2000) *Eradicating Poverty in Uganda: Government Strategy and Operational Framework*, Kampala: Ministries of Agriculture, Animal Industry & Fisheries; and Finance Planning and Economic Development.

Polak, P. (2008) *Out of Poverty – What Works When Traditional Approaches Fail*, San Francisco, CA: Berret-Koehler Publishers.

Polak, P. and Yoder, R. (2006) 'Creating wealth from groundwater for dollar-a-day farmers: Where the silent revolution and the four revolutions to end rural poverty meet', *Hydrogeology Journal*, 14: 424–432.

Poncet J., Kuper M. and Chiche J. (2010) 'Wandering off the paths of planned innovation: The role of formal and informal intermediaries in a large-scale irrigation scheme in Morocco', *Agricultural Systems*, 103: 171–179.

Postel, S., Polak, P., Gonzales, F. and Keller. J. (2001) 'Drip irrigation for small farmers: A new initiative to alleviate hunger and poverty', *Water International*, 26(1): 3–13.

Prakash, A. and Kollman, K. L. (2003) 'Biopolitics in the EU and the U.S.: A race to the bottom or convergence to the top?', *International Studies Quarterly*, 47: 617–641.

Pray, C. E. and Naseem, A. (2007) 'Supplying crop biotechnology to the poor: Opportunities and constraints', *Journal of Development Studies*, 43(1): 192–217.

Pretty, J., Adams, B., Berkes, F., de Athayde, S. F., Dudley, N., Hunn, E., Maffi, L., Milton, K., Rapport, D., Robbins, P., Sterling, E., Stolton, S., Tsing, A., Vintinnerk, E. and Pilgrim, S. (2009) 'The intersections of biological diversity and cultural diversity: Towards integration', *Conservation and Society*, 7(2): 100–112.

Prigogine, I and Stengers, I. (1984) *Order Out of Chaos: Man's New Dialogue with Nature*, New York: Bantam Books.

Quijano, A. (2007) 'Coloniality and modernity/rationality', *Cultural Studies*, 21: 168–178.

RAC (2012) *Why Invest in Orange-fleshed Sweetpotato in Tanzania?*, Policy Brief, Nairobi: International Potato Center.

Ramalingam, B. (2013) *Aid on the Edge of Chaos: Rethinking International Cooperation in a Complex World*, Oxford: Oxford University Press.

Ravetz, J. R. (1999) 'What is post-normal science?' *Futures*, 31: 647–653.

Redman, C. L., Grove, J. M. and Kuby, L. H. (2004) 'Integrating social science into the long-term ecological research (LTER) network: Social dimensions of ecological change and ecological dimensions of social change', *Ecosystems*, 7: 161–171.

Remington, T., Maroko, J., Walsh, S., Omanga, P. and Charles, E. (2002) 'Getting off the seeds–and–tools treadmill with CRS seed vouchers and fairs', *Disasters*, 26: 316–328.

Renkow, M. and Byerlee, D. (2010) 'The impacts of CGIAR research: A review of recent evidence', *Food Policy*, 35(5): 391–402.

Renn, O. (1999) 'A model for an analytic–deliberative process in risk management', *Environmental Science & Technology*, 33: 3049–3055.

Richards, P. (1985) *Indigenous Agricultural Revolution: Ecology and Food Production in West Africa,* London: Unwin Hyman.

Richards, P. (1986) *Coping With Hunger: Hazard and Experiment in an African Rice Farming System,* London: Allen & Unwin.

Richards, P. (1989) 'Agriculture as a performance', in Chambers, R., Pacey, A. and Thrupp, L-A. (eds), *Farmer First: Farmer Innovation and Agricultural Research,* London: Intermediate Technology Publications, pp. 39–42.

Richards, P. (1993) 'Cultivation: Knowledge or performance?', in Hobart, M. (ed.), *An Anthropological Critique of Development: The Growth of Ignorance,* London: Routledge, pp. 61–78.

Richards, P. (1996) 'Culture and community values in the selection and maintenance of African rice', *Indigenous People and Intellectual Property Rights,* Washington, DC: Island Press, pp. 209–229.

Richards, P. and Diemer, G. (1996) 'Agrarian technologies as socio-technical hybrids. Food crop improvement and management of land and water in sub-Saharan Africa', *Bulletin de l'APAD,* 11, available at: https://apad.revues.org/pdf/641 (accessed 15 February 2017).

Richards, P., de Bruin-Hoekzema, M., Hughes, S. G., Kudadjie-Freeman, C., Offei, S., Struik, P. and Zannou, A. (2008) 'Seed systems for African food security: Linking molecular genetic analysis and cultivator knowledge in West Africa', *International Journal of Technology Management,* 45: 196–214.

Ricœur, P. (2005) *The Course of Recognition,* Cambridge MA: Harvard University Press.

Rittel, H. J. and Webber, M. (1973) 'Dilemmas in a general theory of planning', *Policy Sciences,* 4: 155–169.

Robbins, P. (2012) *Political Ecology: A Critical Introduction* (2nd edition), Malden, MA: Wiley-Blackwell.

Roberts, R. L. (1996) *Two Worlds of Cotton: Colonialism and the Regional Economy in the French Soudan, 1800–1946,* Stanford, CA: Stanford University Press.

Robinson, E., Temu, A., Waized, B., Ndyetabula, D., Humphrey, J. and Henson, S. (2014) *Policy Options to Enhance Markets for Nutrient-Dense Foods in Tanzania,* Evidence Report 90, Brighton: Institute of Development Studies.

Rogers, E. M. (2003) *Diffusion of Innovations,* New York: Free Press.

Röling, N. G. and Engel, P. G. H. (1990) 'IT from a knowledge systems perspective: Concepts and issues', *Knowledge in Society: The International Journal of Knowledge Transfer,* 3: 6–18.

Romeis, J., McLean, M. A. and Shelton, A. M. (2013) 'When bad science makes good headlines: Bt maize and regulatory bans', *Nature Biotechnology,* 31: 386–387.

Rosa, M. C. (2014) 'Theories of the South: Limits and perspectives of an emergent movement in social sciences', *Current Sociology,* 62: 851–867.

Ross, C. (2014) 'The plantation paradigm: Colonial agronomy, African farmers, and the global cocoa boom, 1870s–1940s', *Journal of Global History,* 9(1): 49–71.

Roux, D. J., Stirzaker, R. J., Breen, C. M., Lefroy, E. C. and Cresswell, H. P. (2010) 'Framework for participative reflection on the accomplishment of transdisciplinary research programs', *Environmental Science & Policy,* 13(8): 733–741.

Ruel, M. T. and Alderman, H. (2013) 'Nutrition-sensitive interventions and programs: How can they help accelerate progress in improving maternal and child nutrition?', *Lancet,* 382(9891): 536–551.

Ruttan, V. W. and Y. Hayami (1973) 'Technology transfer and agricultural development', *Technology and Culture,* 14(2): 119–151.

Rwamigisa P. B., Mangheni M. N. and Birner R. (2012) *The Challenge of Reforming National Agricultural Extension Systems in Africa: The Case of Uganda's Policy Reform Process 1996–2011*, International Conference Proceedings on Innovations in Extension and Advisory Services, held in Nairobi, Kenya 2011.

Rwampororo, R. W., Negra, C. and Kueneman, E. (2016) *Humidtropics Evaluation Report*, Rome, Italy: Independent Evaluation Arrangement (IEA) of the CGIAR.

Ryan, J. (2006) 'International public goods and the CGIAR niche in the R for D continuum: Operationalising concepts', in CGIAR Science Council (2006) *Positioning the CGIAR in the Global Research for Development Continuum*. Rome, Italy: Science Council Secretariat, pp. 1–24.

Ryan, O. (2011) *Chocolate Nations: Living and Dying for Cocoa in West Africa*, New York: Zed Books.

Sabatier, P. A. and Jenkins-Smith, H. C. (eds) (1993) *Policy Change and Learning: An Advocacy Coalition Approach*, Boulder, CO: Westview Press.

Sabatier, P. A. and Pelkey, N. (1987) 'Incorporating multiple actors and guidance instruments into models of regulatory policy-making: An advocacy coalition framework', *Administration and Society*, 19(2): 236–263.

Sagasti, F. and Timmer, V. (2008) 'System-wide review of the CGIAR system. An approach to the CGIAR as a provider of international public goods', Revised version of a draft submitted as a contribution to the work of the Independent Review Panel (IRP) of the CGIAR.

Saltzman, A., Birol, E., Bouis, H. E., Boy, E., De Moura, F. F., Islam, Y. and Pfeiffer, W. H. (2013) 'Biofortification: Progress toward a more nourishing future', *Global Food Security*, 2(1): 9–17.

Salzburg Global Seminar 2008, IDS/Future Agriculture/Salzburg Global Seminars.

Save the Children Tanzania, Sokoine University of Agriculture and PANITA (2012) *Nutrition Policy Mapping for Tanzania*, Dar es Salaam: Save the Children Tanzania.

Sayer, A. (2011) *Why Things Matter to People: Social Science, Values and Ethical Life*, Cambridge: Cambridge University Press.

Sayer, A. (2015) 'Time for moral economy?', *Geoforum*, 65: 91–293.

Scanagri Consulting Company Ltd. (2005) *Mid-Term evaluation of Uganda's National Agricultural Advisory Services (NAADS) program*, Draft Consultancy report submitted to the Government of Uganda by Scanagri Ltd UK.

Schmidt, V. (2014) *Nutrition Security in Tanzania*, Hamburg: Diplomica Verlag.

Schneiberg, M., King, M. and Smith, T. (2008) 'Social movements and organizational form: Cooperative alternatives to corporations in the American insurance, dairy, and grain industries', *American Sociological Review*, 73(4): 635–667.

Schön, D. A. and Rein, M. (1994) *'Frame Reflection', Toward the Resolution of Intractable Policy Controversies*, New York: Basic Books.

Schumacher, E. F. (1973) *Small is Beautiful: Economics as if People Mattered*, New York: Harper.

Schurman, R. and Munro, W. (2014) 'The institutional architecture of the New Green Revolution for Africa', Presentation at the annual meeting of the African Studies Association. Indianapolis, IN USA. 20-23 November.

Schut, M., Klerkx, L. and Leeuwis, C. (2015a) *Rapid Appraisal of Agricultural Innovation Systems (RAAIS). A toolkit for integrated analysis of complex agricultural problems and innovation capacity in agrifood systems*, International Institute of Tropical Agriculture (IITA) and Wageningen University. Available online: http://www.wur.nl/en/article/RAAIS-Toolkit.htm (accessed 1 February 2017).

Schut, M., Klerkx, L., Rodenburg, J., Kayeke, J., Hinnou, L. C., Raboanarielina, C., Adegbola, P. Y., van Ast, A. and Bastiaans, L. (2015b) 'RAAIS: Rapid Appraisal of Agricultural Innovation Systems (Part I). A diagnostic tool for integrated analysis of complex problems and innovation capacity', *Agricultural Systems*, 132: 1–11.

Schut, M., Klerkx, L., Sartas, M., Lamers, D., Campbell, M., Ogbonna, I., Kaushik, P., Atta-Krah, K. and Leeuwis, C. (2016a) 'Innovation platforms: Experiences with their institutional embedding in agricultural research for development', *Experimental Agriculture*, 52(4): 537–561.

Schut, M., van Asten, P., Okafor, C., Hicintuka, C., Mapatano, S., Nabahungu, N.L., Kagabo, D., Muchunguzi, P., Njukwe, E., Dontsop-Nguezet, P. M., Sartas, M. and Vanlauwe, B. (2016b) 'Sustainable intensification of agricultural systems in the Central African Highlands: The need for institutional innovation', *Agricultural Systems*, 145: 165–176.

Schut, M., van Paassen, A., Leeuwis, C. and Klerkx, L. (2014) 'Towards dynamic research configurations: A framework for reflection on the contribution of research to policy and innovation processes', *Science and Public Policy*, 41(2): 207–218.

Scoones, I. and Thompson, J. (eds) (1994) *Beyond Farmer First: Rural People's Knowledge, Agricultural Research and Extension Practice*, London: Intermediate Technology Publications.

Scoones, I. and Thompson, J. (2011) 'The politics of seed in Africa's green revolution: Alternative narratives and competing pathways', *IDS Bulletin*, 42: 1–23.

Scoones, I., Amanor, K., Favareto, A. and Qi, G. (2016) 'A new politics of development cooperation? Chinese and Brazilian engagements in African agriculture', *World Development*, 81: 1–12.

Scott, J. C. (1985) *Weapons of the Weak: Everyday Forms of Peasant Resistance*. New Haven: Yale University Press.

Scott, J. C. (1998) *Seeing Like a State: How Certain Schemes to Improve the Human Condition Have Failed*, New Haven: Yale University Press.

Scott, J. C. (2008) *Weapons of the Weak: Everyday Forms of Peasant Resistance*, New Haven NJ: Yale University Press.

Scott, J. C. (2009) 'Vernaculars cross-dressed as universals: Globalization as North Atlantic hegemony', *Macalester International*, 24(7): 3–29.

Seckler D. (1996) *The New Era of Water Resources Management: From 'Dry' to 'Wet' Water Savings*, International Irrigation Management Research (IIMI) Report No. 1, Colombo, Sri Lanka: IMMI.

Semana, A. R. (2002) 'Agricultural extension services at crossroads: Present dilemma and possible solutions for future in Uganda', paper presented at the Codesria-IFS Sustainable Agriculture Initiative Workshop, Kampala, Uganda. 15–16 December 2002.

Sen, A. (1981) *Poverty and Famines: An Essay on Entitlements and Deprivation*, Oxford: Clarendon Press.

Sen, A. (2001) *Development as Freedom*, Oxford: Oxford University Press.

Sen, A. (2009) *The Idea of Justice*, Cambridge MA: Harvard University Press.

Sen, D. (2015) *How Smallholder Farmers in Uttarakhand Reworked the System of Rice Intensification: Innovations from Sociotechnical Interactions in Fields and Villages*, PhD thesis, Wageningen University.

Séralini, G.-E., Clair, E., Mesnage, R., Gress, S., Defarge, N., Malatesta, M., Hennequin, D. and de Vendômois, J.S. (2012) 'Long term toxicity of a Roundup herbicide and a Roundup-tolerant genetically modified maize', *Food and Chemical Toxicology*, 50: 4221–4231.

Shankland, A. and Gonçalves, E. (2016) 'Imagining agricultural development in South-South cooperation: The contestation and transformation of ProSAVANA', *World Development*, 81: 35–46.

Shapin, S. (1998) 'Placing the view from nowhere: Historical and sociological problems in the location of science', *Transactions of the Institute of British Geographers*, 23(1): 5–12.

Shen, X. (2013) *Private Chinese Investment in Africa: Myths and Realities,* Policy Research Working Paper 6311. Washington, DC: World Bank.

Shepherd, A. and Onumah, G. E. (1997) *Liberalised Agricultural Markets in Ghana: The Role and Capacity of Government. The Role of Government in Adjusting Economies,* Paper 12, Development Administration Group, Birmingham: University of Birmingham.

Shiva, V. (1991) *The Violence of the Green Revolution: Third World Agriculture, Ecology, and Politics,* London: Zed Books.

Sillitoe, P. (2010) 'Trust in development: Some implications of knowing in indigenous knowledge', *Journal of the Royal Anthropological Institute*, 16: 12–30.

Silva, J. V., Reidsma, P., Laborte, A. G. and Van Ittersum, M. K. (2016) 'Explaining rice yields and yield gaps in Central Luzon, Philippines: An application of stochastic frontier analysis and crop modelling', *European Journal of Agronomy*, 82(Part B): 223–241, doi:10.1016/j.eja.2016.06.017.

Sindi, K. and Wambugu, S. (2012) *Going-to-Scale with Sweetpotato Vines Distribution in Tanzania,* Marando Bora Baseline Study, Milestone Report OB3BMS2.1C1, Nairobi: International Potato Centre.

Smit, B. and Wandel, J. (2006) 'Adaptation, adaptive capacity and vulnerability', *Global Environmental Change*, 16: 282–292.

Sperling, L. and McGuire, S. (2010) 'Understanding and strengthening informal seed markets', *Experimental Agriculture*, 46: 119–136.

Sperling, L., Cooper, H. D. and Remington, T. (2008) 'Moving towards more effective seed aid', *Journal of Development Studies*, 44(4): 586–612.

Spielman, D. J. and Pandya-Lorch, R. (2009) *Millions Fed: Proven Successes in Agricultural Development*, Washington, DC: IFPRI.

Stathers, T., Mkumbira, J., Low, J. Tagwirey, J. Munyua, H., Mbabu, A. and Mulongo, G. (2015) *Orange-fleshed Sweetpotato (OFSP) Investment Guide*, Nairobi: International Potato Centre.

Stirling, A. (1999) 'The appraisal of sustainability: Some problems and possible responses', *Local Environment*, 4: 111–135.

Stirling, A. (2008) '"Opening up" and "closing down" power, participation, and pluralism in the social appraisal of technology', *Science, Technology & Human Values*, 33: 262–294.

Stone, G. D. (2011) 'Field versus farm in Warangal: Bt cotton, higher yields, and larger questions', *World Development*, 39(3): 387–398.

Stone, G. D. (2016) 'Towards a general theory of agricultural knowledge production: Environmental, social and didactic learning', *Culture, Agriculture, Food and Environment (CAFE)*, 38(1): 5–17.

Stone, G. D. and Glover, D. (2011) 'Genetically modified crops and the 'food crisis': Discourse and material impacts', *Development in Practice*, 21(4–5): 509–516.

Stoop, W. (2011) 'The scientific case for system of rice intensification and its relevance for sustainable crop intensification', *International Journal of Agricultural Sustainability*, 9(3): 443–455.

Stoop, W. and van Walsum, E. (2013) 'SRI - A grassroots revolution', *Farming Matters* [formerly LEISA Magazine], 29(1): 8–9.

Stoop, W., Uphoff, N. and Kassam, A. (2002) 'A review of agricultural research issues raised by the system of rice intensification (SRI) from Madagascar: Opportunities for improving farming systems for resource-poor farmers', *Agricultural Systems*, 71(3): 249–274.

Struik, P. C., Klerkx, L. and Hounkonnou, D. (2014) 'Unravelling institutional determinants affecting change in agriculture in West Africa', *International Journal of Agricultural Sustainability*, 12(3): 370–382.

Stuart, A. M., Pame, A. R. P., Silva, J. V., Dikitanan, R. C., Rutsaert, P., Malabayabas, A. J. B., Lampayan, R. M., Radanielson, A. M. and Singleton, G. R. (2016) 'Yield gaps in rice-based farming systems: Insights from local studies and prospects for future analysis', *Field Crops Research*, 194: 43–56, doi:10.1016/j.fcr.2016.04.039.

Suchman, L. A. (1987) *Plans and Situated Actions: The Problem of Human-Machine Communication*, Cambridge, UK/New York, USA/Melbourne, AU: Cambridge University Press.

Sumberg, J. (2012) 'Mind the (yield) gap(s)', *Food Security*, 4: 509–518.

Sumberg, J. and Okali, C. (1997) *Farmers' Experiments: Creating Local Knowledge*, Boulder, CO: Lynne Rienner Publishers, Inc.

Sumberg, J. and Thompson, J. (eds) (2012) *Contested Agronomy: Agricultural Research in a Changing World*, London: Routledge.

Sumberg, J., Irving, R., Adams, E. and Thompson, J. (2012) 'Success making and success stories: Agronomic research in the spotlight', in Sumberg, J. and Thompson, J. (eds), *Contested Agronomy: Agricultural Research in a Changing World*, London: Routledge.

Sumberg, J., Keeney, D. and Dempsey, B. (2012) 'Public agronomy: Norman Borlaug as "brand hero" for the Green Revolution', *Journal of Development Studies*, 48(11): 1587–1600.

Sumberg, J., Thompson, J. and Woodhouse, P. (2012) 'Contested agronomy: Agricultural research in a changing world', in Sumberg, J. and Thompson, J. (eds), *Contested Agronomy: Agricultural Research in a Changing World*, London: Routledge.

Sumberg, J., Thompson, J. and Woodhouse, P. (2013) 'Why agronomy in the developing world has become contentious', *Agriculture and Human Values*, 30(1): 71–83.

Surridge, C. (2004) 'Feast or famine?', *Nature*, 428(6981): 360–361.

Takei, E. (1984) 'Variation and geographical distribution of cultivated plants in the Southern Sudan', in Sakamoto, K. (ed.), *Agriculture and Land Utilization in the Eastern Zaire and the Southern Sudan*, Kyoto, Japan: Department of Agriculture and Forestry Economics, Faculty of Agriculture, Kyoto University, pp. 53–76.

Tanzania National Business Council (2009) *Kilimo Kwanza*, Dar es Salaam: Tanzania National Business Council.

Taylor, C. (1994) 'The politics of recognition', in Gutmann, A. (ed.), *Multiculturalism: Examining the Politics of Recognition*, Princeton NJ: Princeton University Press, pp. 25–73.

Taylor, F. W. (1947) *Scientific Management*, 3rd edn, New York: Harper & Brothers.

Taylor, I. (1998) 'China's foreign policy towards Africa in the 1990s', *The Journal of Modern African Studies*, 36(3): 433–460.

Temu, A., Waized, B., Ndyetabula, D., Robinson, E., Humphrey, J. and Henson, S. (2014) *Mapping Value Chains for Nutrient-dense Foods in Tanzania*, IDS Evidence Report 76, Brighton: IDS.

Teshome, A., Fahrig, L., Torrance, J. K., Lambert, J., Arnason, T. and Baum, B. (1999) 'Maintenance of sorghum (Sorghum bicolor, Poaceae) landrace diversity by farmers' selection in Ethiopia', *Economic Botany*, 53: 79–88.

Thiele, G. (1999) 'Informal potato seed systems in the Andes: Why are they important and what should we do with them?', *World Development*, 27: 83–99.

Thiele, G., Fliert, E. van der and Campilan, D. (2001) 'What happened to participatory research at the International Potato Center?', *Agriculture and Human Values*, 18(4): 429–446.

Thompson, J. and Scoones, I. (1994) 'Challenging the populist perspective: Rural people's knowledge, agricultural research, and extension practice', *Agriculture and Human Values* 11(2–3): 58–76.

Thompson, J. and Scoones, I. (2009) 'Addressing the dynamics of agri-food systems: An emerging agenda for social science research', *Environmental Science & Policy*, 12: 386–397.

Thompson, J. and Sumberg, J. (2012) 'Nullius in verba: Contestation, pathways and political agronomy', in Sumberg, J., Thompson, J. (eds), *Contested Agronomy: Agricultural Research in a Changing World*, London: Routledge.

Thompson, M. and Warburton, M. (1985) 'Decision making under contradictory certainties: How to save the Himalayas when you can't find out what's wrong with them', *Journal of Applied Systems Analysis*, 12: 3–34.

Thompson, S. (2006) *The Political Theory of Recognition: A Critical Introduction*, Cambridge UK & Malden MA: Polity Press.

Toenniessen, G., Adesina, A. and DeVries, J. (2008) 'Building an alliance for a green revolution in Africa', *Annals of the New York Academy of Sciences*, 1136 (1): 233–42.

Tomlinson, I. (2013) 'Doubling food production to feed the 9 billion: A critical perspective on a key discourse of food security in the UK', *Journal of Rural Studies*, 29: 81–90.

Tripp, R. (ed.) (1997) *New Seed and Old Laws: Regulatory Reform and the Diversification of National Seed Systems*, Rugby: Intermediate Technology Publications Ltd (ITP).

Tsing, A. (2000) 'Inside the economy of appearances', *Public Culture* 12(1): 115–144.

Tugendhat, H. and Alemu, D. (2016) 'Chinese agricultural training courses for African officials: Between power and partnerships', *World Development*, 81: 71–81.

Turner II, B.L. and Robbins, P. (2008) 'Land-change science and political ecology: Similarities, differences, and implications for sustainability science', *Annual Review of Environment and Resources*, 33: 295–316.

Uganda Bureau of Statistics (2005) *The Uganda National Household Survey 2005 Statistical Abstract*, Kampala, Uganda: Statistics House.

Uganda Bureau of Statistics (2008) *The National Service Delivery Survey 2008 Statistical Abstract* Kampala, Uganda: Statistics House.

UK Foresight (2011) *The Future of Food and Farming: Challenges and Choices for Global Sustainability, Final Project Report*, London: UK Government Office for Science.

UN (2015) *Agricultural Technology for Development: Seventieth Session of the UN General Assembly. Item 20 of the Provisional Agenda: Sustainable Development*, New York: United Nations.

UNICEF (2007) *Vitamin A Supplementation: A Decade of Progress*, New York: UNICEF.

UNICEF (2013) *Improving Child Nutrition: The Achievable Imperative for Global Progress*, available at: http://www.who.int/pmnch/media/news/2013/20130416_unicef_factsheet.pdf (accessed 1 February 2017).

UNICEF (2015) UNICEF DATA: Monitoring the situation of Women and Children. Country Statistics for Malnutrition retrieved June 2016. Available at: http://data.unicef.org/nutrition/malnutrition.html (accessed 1 February 2017).

United Republic of Tanzania (2006) *National Sample Census of Agriculture 2002/2003. Volume IV: Gender Profile of Smallholder Rural Agriculture Population in Tanzania Mainland*, Dar es Salaam: Ministry of Agriculture.

United Republic of Tanzania (2011) *National Nutrition Strategy. JULY 2011/12 – JUNE 2015/16*, Dar es Salaam: Ministry of Health and Social Welfare.

United Republic of Tanzania (2014) *National Nutrition Survey 2014,* Dar es Salaam: Ministry of Health and Social Welfare.

Uphoff, N. (1999) 'Agroecological implications of the System of Rice Intensification (SRI) in Madagascar', *Environment, Development and Sustainability*, 1(3–4): 297–313.

Uphoff, N. (2002) Opportunities for raising yields by changing management practices: The System of Rice Intensification in Madagascar', in Uphoff, N. (ed.), *Agroecological Innovations: Increasing Food Production with Participatory Development*, London: Earthscan, pp. 145–161.

Uphoff, N. (2007) 'Agroecological alternatives: Capitalising on existing genetic potentials', *Journal of Development Studies*, 43(1): 218–236.

Uphoff, N., Kassam, A. and Harwood, R. (2011) 'SRI as a methodology for raising crop and water productivity: Productive adaptations in rice agronomy and irrigation water management', *Paddy and Water Environment*, 9(1): 3–11.

Uphoff, N., Kassam, A. and Stoop, W. (2008) 'A critical assessment of a desk study comparing crop production systems: The example of the "system of rice intensification" versus "best management practice"', *Field Crops Research*, 108(1): 109–114.

van Beusekom, M. M. (1997) 'Colonisation indigène: French rural development ideology at the Office du Niger, 1920-1940', *The International Journal of African Historical Studies*, 30(2): 299–323.

van Damme, J., Ansoms, A. and Baret, P. V. (2014) 'Agricultural innovation from above and from below: Confrontation and integration on Rwanda's Hills', *African Affairs*, 113(450): 108–127.

van de Walle, N. (2007) *African Economies and the Politics of Permanent Crisis, 1979–1999: Political Economy of Institutions and Decisions*, Cambridge: Cambridge University Press.

van den Bold, M., Quisumbing, A. R. and Gillespie, S. (2013) 'Women's empowerment and nutrition: An evidence review', IFPRI Discussion Paper 01294, Washington, DC: IFPRI.

van der Kooij, S., Zwarteveen, M., Kuper, M. and Errah M. (2013) 'The efficiency of drip irrigation unpacked', *Agricultural and Water Management*, 123: 103–110.

van der Ploeg, J. D. (2010) 'The peasantries of the twenty-first century: The commoditisation debate revisited', *The Journal of Peasant Studies*, 37(1): 1–30.

van Ittersum, M. K., Cassman, K. G., Grassini, P., Wolf, J., Tittonell, P. and Hochman, Z. (2013) 'Yield gap analysis with local to global relevance – a review', *Field Crops Research*, 143: 4–17.

van Jaarsveld, P. J., Faber, M., Tanumihardjo, S. A., Nestel, P., Lombard, C. J. and Spinnler Benadé, A. J. (2005) 'ß-carotene-rich orange-fleshed sweet potato improves the vitamin A status of primary school children assessed with the modified-relative-dose-response test', *American Journal of Clinical Nutrition*, 81(5): 1080–1087.

van Noordwijk, M. and Brussaard, L. (2014) 'Minimizing the ecological footprint of food: Closing yield and efficiency gaps simultaneously?', *Current Opinion In Environmental Sustainability*, 8: 62–70.

van Noordwijk, M., Cadisch, G. and Ong, C. K. (eds) (2004) *Below-Ground Interactions in Tropical Agroecosystems. Concepts and Models with Multiple Plant Components*, Wallingford: CAB International.

van Vliet, J. A., Schut, A. G. T., Reidsma, P., Descheemaeker, K., Slingerland, M., van de Ven, G. W. J. and Giller, K. E. (2015) 'De-mystifying family farming: Features, diversity and trends across the globe', *Global Food Security*, 5: 11–18.

van Zwanenberg, P., Ely, A. and Smith, A. (2008) *Rethinking Regulation: International Harmonisation and Local Realities*, STEPS Working Paper 12, Brighton: STEPS Centre.

van Zwanenberg, P., Ely, A., Smith, A., Chuanbo, C., Shijun, D., Fazio, M.-E. and Goldberg, L. (2011) 'Regulatory harmonization and agricultural biotechnology in Argentina and China: Critical assessment of state-centered and decentered approaches', *Regulation & Governance*, 5: 166–186.

Vanlauwe, B., Coe, R. and Giller, K. E. (2016) 'Beyond averages: New approaches to understand heterogeneity and risk of technology success or failure in smallholder farming', *Experimental Agriculture*, doi:10.1017/S0014479716000193.

Vanloqueren, G. and Baret, P. V. (2009) 'How agricultural research systems shape a technological regime that develops genetic engineering but locks out agroecological innovations', *Research Policy*, 38(6): 971–983.

Venot, J. P. (2016) 'A success of some sort: Drip irrigation social enterprises and drip irrigation in the developing world', *World Development*, 79: 69–81.

Venot, J. P., Zwarteveen, M., Kuper, M., Boesveld, H., Bossenbroek, L., van der Kooij, S., Wanvoeke, J., Benouniche, M., Errahj, M., de Fraiture, C. and Verma. S. (2014) 'Beyond the promises of technology: A review of the discourses and actors who make drip irrigation', *Irrigation and Drainage*, 63(2): 186–194.

Vernooy, R., Shrestha, P. and Sthapit, B. (2015) *Community Seed Banks: Origins, Evolution and Prospects*, London: Routledge.

Virchow, D. (2013) *Nutrition-Sensitive Agriculture: A Pillar of Improved Nutrition and Better Health*, Food Security Center, University of Hohenheim.

Virchow, D. and von Braun, J. (2001) 'Dresden declaration: Towards a global system for agricultural research for development', in Virchow, D., von Braun, J. (eds), *Villages in the Future*, Berlin: Springer, pp. 321–322.

Visvanathan, S. (1997) *A Carnival for Science: Essays on Science, Technology, and Development*, Oxford: Oxford University Press.

Visvanathan, S. (2005) 'Knowledge, justice and democracy', in Leach, M., Scoones, I., Wynne, B. (eds), *Science and Citizens: Globalization and the Challenge of Engagement*, London & New York: Zed Books.

von Braun, J. (2011) 'Addressing the food crisis: Governance, market functioning and investment in public goods', *Food Security*, 1(1): 9–15.

von Kaufmann, R. (2007) 'Integrated Agricultural Research for Development: Contributing to the Comprehensive Africa Agricultural Development Programme (IAR4D in CAADP)', in Bationo, A., Waswa, B., Kihara, J. and Kimetu, J. (eds), *Advances in Integrated Soil Fertility Management in Sub-Saharan Africa: Challenges and Opportunities*, Berlin: Springer Netherlands.

Waized, B., Ndyetabula, D., Temu, A., Robinson, E. and Henson, S. (2015) *Reducing Hunger and Nutrition: Promoting Biofortified Crops for Nutrition: Lesson from Orange-fleshed Sweet Potato (OFSP) in Tanzania*, Brighton: Institute of Development Studies.

Wakhungu, J. W. and Wafula, D. (2004) *Introducing Bt. cotton: Policy Lessons for Smallholder Farmers in Kenya*, Nairobi: African Centre for Technology Studies.

Wambugu, F. (1999) 'Why Africa needs agricultural biotech', *Nature*, 400: 15–16.

Wanvoeke, J., Venot, J. P., de Fraiture, C. and Zwarteveen, M. (2015) 'Smallholder drip irrigation in Burkina Faso: The role of development brokers', *Journal of Development Studies*, 52(7): 1019–1033.

Wanvoeke, J., Venot, J. P., Zwarteveen, M. and de Fraiture, C. (2016) 'Farmers' logics in engaging with projects promoting drip irrigation kits in Burkina Faso', *Society & Natural Resources*, 29(9): 1095–1109.

Wattnem, T. (2016) 'Seed laws, certification and standardization: Outlawing informal seed systems in the Global South', *The Journal of Peasant Studies*, 3(4): 850–867, doi:10.1080/03066150.2015.1130702.

Waziri, M. (2013) *Cassava and Sweet Potato Value Chains in Mvomero and Kongwa Districts in Tanzania*. Master's Dissertation, Sokoine University of Agriculture, Morogoro, Tanzania.

Weale, A. (1992) *The New Politics of Pollution*, Manchester: Manchester University Press.

Webster, J. (1991) 'Advanced manufacturing technologies: Work organisation and social relations crystallised', in Law, J. (ed.), *A Sociology of Monsters: Essays on Power, Technology and Domination*, London: Routledge, pp. 192–222.

Westengen, O. T. and Brysting, A. K. (2014) 'Crop adaptation to climate change in the semi-arid zone in Tanzania: The role of genetic resources and seed systems', *Agriculture & Food Security*, 3: 3 (doi:10.1186/2048-7010-3-3).

Westengen, O. T., Okongo, M. A., Onek, L., Berg, T., Upadhyaya, H., Birkeland, S., Khalsa, S. D. K., Ring, K. H., Stenseth, N. C. and Brysting, A. K. (2014a) 'Ethnolinguistic structuring of sorghum genetic diversity in Africa and the role of local seed systems', *Proceedings of the National Academy of Sciences*, 111: 14100–14105.

Westengen, O. T., Ring, K. H., Berg, P. R. and Brysting, A. K. (2014b) 'Modern maize varieties going local in the semi-arid zone in Tanzania', *BMC evolutionary biology*, 14: 1.

Wezel, A., Bellon, S., Dore, T., Francis, C., Vallod, D. and David, C. (2009) 'Agroecology as a science, a movement and a practice. A review', *Agronomy for Sustainable Development*, 29: 503–515.

Whitfield, S. (2016) *Adapting to Climate Uncertainty in African Agriculture: Narratives and Knowledge Politics*, London: Routledge.

Wiggins, S. (2009) 'Can the smallholder model deliver poverty reduction and food security for a rapidly growing population in Africa?', *Expert Meeting in How to Feed the World in 2050*, Rome: Economic and Social Development Department, FAO.

Willer, H and Kilcher, L. (eds) (2016) *The World of Organic Agriculture. Statistics and Emerging Trends 2016*, Bonn and Frick: IFOAM and FiBL.

Wittman, H. (2009) 'Reworking the metabolic rift: La Vía Campesina, agrarian citizenship, and food sovereignty', *The Journal of Peasant Studies*, 36: 805–826.

Wood, D. and Lenné, J. (2001) 'Nature's fields: A neglected model for increasing food production', *Outlook on Agriculture*, 30(3): 161–170.

Woodward, A. D. (2009) 'The impact of US subsidies on West African cotton production', in Pinstrup-Anderson, P. and Cheung, F. (eds), *Studies in Food Policy for Developing Countries: Institutions and International Trade Policies*, Ithaca: Cornel University Press, pp. 195–204.

Woolgar, S. (1991) 'Configuring the user: The case of usability trials', in Law, J. (ed.), *A Sociology of Monsters: Essays on Power, Technology and Domination*, London: Routledge, pp. 57–99.

Woomer, P. L. and Swift, M. J. (eds) (1994) *The Biological Management of Tropical Soil Fertility*, Chichester, UK: John Wiley.

World Bank (2007) *World Development Report 2008: Agriculture for Development*, Washington, DC: World Bank.

World Bank (2010a) *Agricultural Technology Agribusiness and Advisory Services*. Project Appraisal Document, Washington, DC: World Bank.

World Bank (2010b) *Implementation Completion and Results Report (IDA 34630) on an IDA Credit in the amount of SDR 35.3 Million (USD 45.0 Million equivalent) to the Republic of Uganda for the National Agricultural Advisory Services Project, Report No. ICR00001421,*

World Bank, Agriculture and Rural Development Unit, Sustainable Department, Africa Region.

Xu, X., Li, X., Qu, G., Tang, L. and Mukwezereza, L. (2016) 'Science, technology, and the politics of knowledge: The case of China's agricultural technology demonstration centers in Africa', *World Development*, 81: 82–91.

Zhang, Q. F. and Donaldson, J. A. (2008) 'The rise of Agrarian capitalism with Chinese characteristics: Agricultural modernisation, agribusiness and collective land rights', *The China Journal*, 60: 25–47.

Zurn, C. (2015) *Axel Honneth: A Critical Theory of the Social*, Cambridge UK & Malden MA: Polity Press.

INDEX

Page numbers in *italics* denote figures, those in **bold** denote tables.

For Product Safety Concerns and Information please contact our EU
representative GPSR@taylorandfrancis.com
Taylor & Francis Verlag GmbH, Kaufingerstraße 24, 80331 München, Germany

9 781138 240315